Ready, Set, Weld!

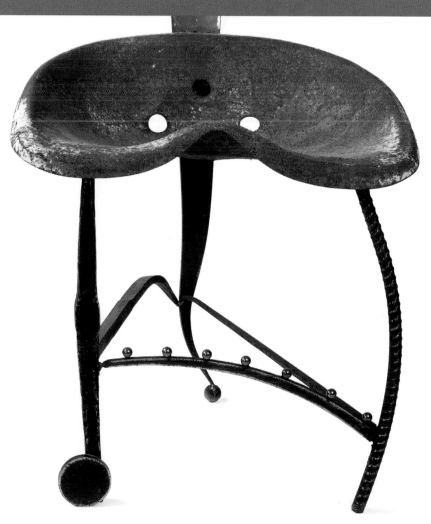

Ready, Set, Weld!

Beginner-Friendly Projects for the Home & Garden

Kimberli Matin

LARK BOOKS

A Division of Sterling Publishing Co., Inc.

New York / London

Senior Editor: **Terry Taylor**
Editor: **Amanda Carestio**
Art Director: **Kristi Pfeffer**
Art Production: **Carol Morse**
Photographer: **Steve Mann**
Cover Designer: **Susan McBride**

Library of Congress Cataloging-in-Publication Data

Matin, Kimberli.
  Ready, set, weld! : beginner-friendly projects for the home & garden /
Kimberli Matin. – 1st ed.
     p. cm.
  Includes index.
  ISBN 978-1-60059-262-1 (pbk. : alk. paper)
  1. Welding. I. Title.
  TT211.M37 2008
  671.5'2–dc22

                                        2008033519

10 9 8 7 6 5 4 3 2 1

First Edition

Published by Lark Books, A Division of
Sterling Publishing Co., Inc.
387 Park Avenue South, New York, NY 10016

Text © 2009, Kimberli Matin
Photography © 2009, Lark Books unless otherwise specified

Distributed in Canada by Sterling Publishing,
c/o Canadian Manda Group, 165 Dufferin Street
Toronto, Ontario, Canada M6K 3H6

Distributed in the United Kingdom by GMC Distribution
Services,Castle Place, 166 High Street, Lewes, East Sussex,
England BN7 1XU

Distributed in Australia by Capricorn Link (Australia) Pty Ltd.,
P.O. Box 704, Windsor, NSW 2756 Australia

The written instructions, photographs, designs, patterns, and projects
in this volume are intended for the personal use of the reader and
may be reproduced for that purpose only. Any other use, especially
commercial use, is forbidden under law without written permission
of the copyright holder.

Every effort has been made to ensure that all the information in
this book is accurate. However, due to differing conditions, tools,
and individual skills, the publisher cannot be responsible for any
injuries, losses, and other damages that may result from the use of
the information in this book.

If you have questions or comments about this book, please contact:
Lark Books
67 Broadway
Asheville, NC 28801
828-253-0467

Manufactured in China

ISBN 13: 978-1-60059-262-1

For information about custom editions, special sales, premium
and corporate purchases, please contact Sterling Special Sales
Department at 800-805-5489 or specialsales@sterlingpub.com.

DEDICATION

I dedicate this book to my beautiful mother, a fellow artist
who instilled in me the idea that I could do it!

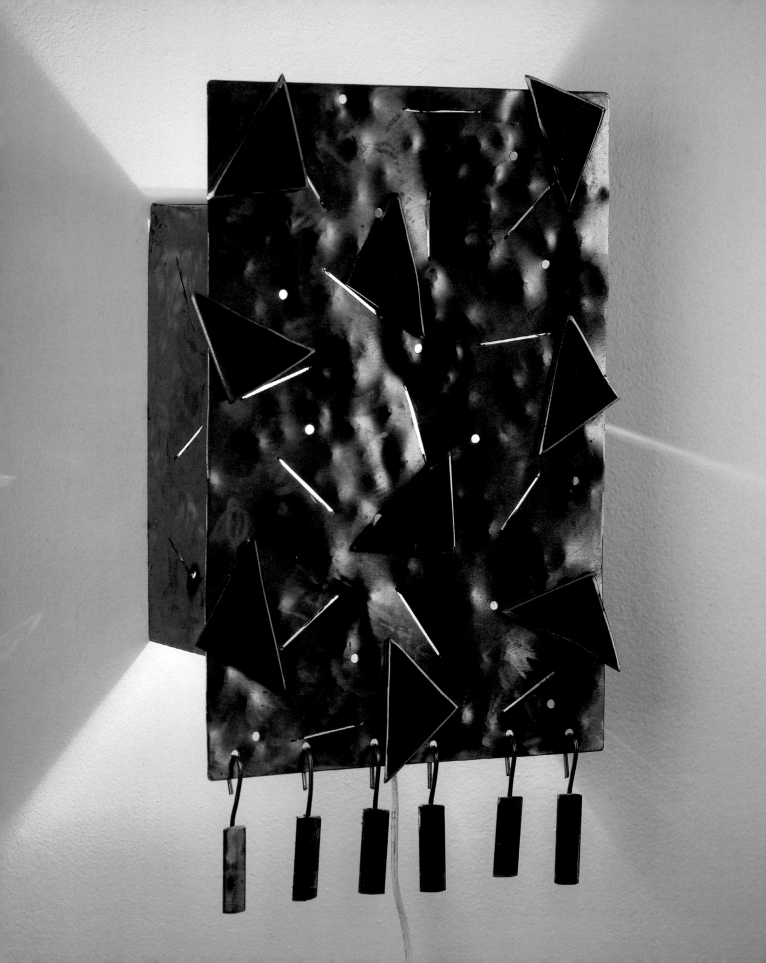

▲ Table of Contents

Introduction

I'll admit it: I was scared the first time I pulled the trigger of a MIG welding gun. Everything went black, there was a bright flash, sparks flew out, and I heard a loud crackling sound. But I got over that initial fear quickly, and you will, too. In fact, you'll soon discover, as I did, that operating a MIG welder is basically like using a large glue gun for metal.

Like many women, I was not raised around power tools. And I certainly wasn't encouraged to work with steel. But while watching a neighbor weld one day, I realized a) that welding wasn't as hard as I'd thought, and b) that steel is both versatile and beautiful. I fell in love.

I purchased my own MIG welder and looked into what the local welding classes had to offer—but I didn't need to learn arc or TIG welding. I wasn't setting out to build a suspension bridge; I just wanted to make picture frames or perhaps assemble a small table. That's when I decided to stop being intimidated by a simple process and started working with my MIG welder and a few simple tools. Since I couldn't keep the names of my tools straight, I just made up my own names for them. Little Spark (my welder) and Minnie Mouse (my angle grinder) became my friends. Welding tools tend not to come in pinks and purples, so I decorated mine to suit my fancy. I've painted flowers on my gloves and even glued rhinestones to my welding helmet to match my earrings. Then I started having fun and getting creative with welding.

One day when I was bending metal for the back of a chair, I had a lightbulb moment. I could simplify the process by using premade parts for railings and gates to create things like benches and mirrors. Available in a variety of inspiring shapes in catalogs and online, purchasing shapes instead of making them myself would allow me to make all sorts of things out of architectural pieces. Many of the things I created that way are in this book as projects.

Lightbulb number two came on during my very first visit to a scrapyard. I looked down and saw a piece of bent scrap metal that looked like a face. Then I found parts for the eyes, nose, and hair, and welded together one of my most cherished pieces (it hangs near the front door and holds my mail, see photo on page 29). Nothing massages your creative muscle more than the possibilities you'll discover in the mountains of scrap lying around a junkyard.

The Basics section will show you everything you need to know to complete the projects, from cutting metal and making your very first weld to adding creative textures and finishing your piece.

After that, you can start out with easy projects and work your way up to the more challenging ones. Experiment with your own creative ideas right from the start; even the beginning projects hold unlimited creative possibilities. If you begin with the pair of bookends (page 42) made with metal from your local home improvement store, you don't have to copy the squiggles that I created on the surface—think of other ways to make decorative lines, shapes, or patterns. Or use your angle grinder to give the steel a brushed finish.

Let these projects serve as both your introduction to welding and inspiration for your own creative adventures. Once you've mastered simple welds, you'll be ready to come up with your own designs. Look at the instructions for Table for Two (page 82), made with parts you can order from architectural catalogs, and Harvest Table (page 95), constructed with found steel parts. Imagine how easy it will be to design your own table, possibly even combining found *and* purchased parts! Before you know it, you'll be driving to your local junkyard for scrap pieces to add to garden stakes.

Trust me when I say that you can do this. Shed your fears and preconceived notions, and play with steel! It's easier than you think. If a person who used to get nervous when she chipped her manicure can do it, you can, too.

The Basics

Technically, welding is a process that joins or blends two pieces of metal through the use of heat and with or without the application of pressure. It's a simple process that works on simple principles, but it can seem a little intimidating. Most often, a weld is created by using the flow of electricity to heat and melt the edges of two pieces of metal. Filler material is then added between the metal pieces to create a weld pool, which becomes solid and forms a strong bond when it cools.

A Little History

Welding has a long and industrious history. Joining metals became a popular pursuit in the Bronze and Iron Ages in both Europe and the Middle East. The Middle Ages in particular brought about major advances in forge welding, a process that involves pounding molten metal until bonding occurs, and the technique is still used by some blacksmiths.

Popular welding processes remained largely unchanged until the end of the nineteenth century. Advances in technology and an increased demand for welded products, largely for war materials, brought about an expansion in both the uses for welding and the need for skilled welders. One of the first major advances in popular welding methods, oxyacetylene welding began around 1900. Soon, other methods followed: TIG (GTAW) around the 1920s; MIG (GMAW) around the 1930s; and stick (SMAW or arc) and flux-cored (FCAW) around the 1950s.

While there are several different types of welding (see the sidebar on page 27), we're going to use MIG welding for the projects in this book. Often referred to as wire welding, MIG welding is quite easy, basically requiring only three simple tasks: setting the machine, holding the gun, and pulling the trigger. In this process, a welding gun automatically feeds through wire—which serves as additional base material—along with a shielding gas that keeps the welds smooth by preventing oxidation (thus greatly eliminating splatter that would have to be ground smooth later). Like anything else, MIG welding takes time to master, but the average person can create satisfactory welds with just one hour of experimentation. With a little more practice, you'll be creating welds that are both clean and structurally sound enough for the creative projects that follow.

Please note that the welding techniques covered in this book are meant for decorative use only, like those represented by the types of projects here. If you're interested in working heavy load-bearing projects such as repairing a trailer hitch or building a bridge, I suggest you take a course, usually available through your local community college.

Tools

There are basically two kinds of shop people in the world: those who want every kind of tool available—the more the better—and those that are happy to use whatever is lying around if it does the trick. I tend to fall into the latter category. I do, however, feel it's important that each person work in their own personal style, beginning with the tools they use. Feel free to experiment with different potential tools, even if they're not listed here, to find what works best for you.

The MIG Welder Kit

You can buy a basic MIG welder at most large home improvement stores, but if you're new to welding, head to an industrial supply store for a higher level of customer service and expertise. If you have someone to help you, you could even try searching for supplies online where you might be able to find some very good deals. But keep in mind that all the "buyer beware" warnings apply.

Before you select a machine, examine the box to determine if you need a 220-volt outlet (like the kind used for a clothes dryer) or a 110-volt outlet (the standard household outlet) for your machine. If you only have access to 110-volt outlets, you'll need to purchase a MIG welder that uses this type of outlet.

Your MIG welder comes with:

Power box with wire feed

The box holds the power supply and the wire spool. The welding wire is pulled from

MIG Welder Kit

Power box with wire feed
Welding gun
Work cable with
   grounding clamp
Gas nozzle, gas regulator
   hose, and gas cylinder
Helmet and gloves

the spool by a set of rollers powered by a small motor. The wire then travels through the supply cable and out through the welding gun when you press the trigger.

Welding gun

Sometimes referred to as a torch, the welding gun is connected to the power box through a supply cable. The opening at the end of the gun is where you'll be able to see the weld actually happening.

Often students describe this process as similar to using an extremely effective glue gun.

Work cable with grounding clamp

The grounding clamp is also connected to the power box. By connecting the grounding clamp to the metal material you're welding, an arc is established from the electrode through your workpiece. This creates a flow of electricity that results in the heat necessary for the welding process to occur.

Gas nozzle, gas regulator, hose, and gas cylinder

To get started, you'll have to purchase a gas tank, but refilling it later (see page 20) is much less expensive. An industrial supply shop will have different-sized tanks to choose from. I started out with a small 20-cubic-foot tank and then upgraded to a larger 40-cubic-foot cylinder. Your gas source connects to the power box via a connecting hose. Ask the folks at the supply shop to explain how to connect the hose so you'll be ready when you get back to your shop. The regulator tells you how much gas is in the tank and how much gas is flowing through the hose.

Tips

As you work, the wire feeds through a hole in the welding gun tip. Like wire, tips come in different sizes including .023, .030, and .035 inch (0.6, 0.8, and 0.9 mm), and you'll need to use one that matches your wire size.

Angle Grinder

Your angle grinder, also called a side grinder, is the second most important tool in the shop after your welder. A versatile tool with plenty of finishing possibilities (see page 38), angle grinders come with a variety of attachments that screw into one end. The four most common attachments I use are the grinding disc (which I use for etching designs in the surface of the steel and grinding pieces), cut-off wheel (great for cutting small pieces of rod or flat bar when a band saw is not available), knotted wire cup brush (the perfect attachment for cleaning off the small bits of metal that are produced during welding), and the flap disc. The flap disc is great for transforming steel to a beautiful brushed silver surface, and I use it a lot; depending on what kind of work I am doing, I might need to purchase a new flap disc every month or so. There are a few safety concerns that you should consider (see sidebar on page 13), however, before you operate this very powerful and potentially dangerous piece of shop equipment.

Angle Grinder Kit

Angle grinder
Flap disc
Knotted wire cup brush
Grinding disc
Cut-off wheel

Angle Grinder Safety

Keep these tips in mind when operating your angle grinder.

First, always wear *both* gloves. Gloves will give you some cushion between your tender flesh and the wheel.

Wear a full-face shield to protect your eyes, neck, and face from flying debris. Angle grinders come with a wheel guard, great for added protection and for keeping sparks out of your face. A dust mask, like the ones painters use, can keep you from breathing in dust that is created from grinding, as well as rust and other particles that might be on the surface of your material.

The angle grinder is quite loud, so put in your ear protection before you grind anything. Besides protecting your ears from harm, earplugs will also help to relax you. I find that I am much more relaxed when I am not overwhelmed by the loud noise of the grinder.

Always be alert and aware every second that your angle grinder is running, especially if there are other people in the vicinity. Move very slowly and cautiously. Breathe slowly and deeply. It is easy for the wheel to get stuck at an angle, and if it does, it could "jump" in your hand. Your best plan is to maintain a good consistent grip on the grinder with your feet firmly planted at all times.

Also be aware of hanging clothing that might get caught in the wheel. It's easy to get lost in the process, and a loose shirttail could get caught in the wheel, forcing the grinder in toward your body.

Even after you have been grinding for a while, remember that EVERY SINGLE TIME you turn on your angle grinder you should exercise full attention and absolute focus on what you're doing.

Welding Table

Whether you purchase one or make your own, your welding table provides a stable area for laying out pieces before you weld them together. Most welding tables that you purchase will be adjustable. If you decide to make your own (look online for free plans), make sure the height is set at a comfortable level so you can work without bending over or having the work too high in your face. A good rule of thumb for correct table height is about wrist level.

For a simple homemade table, buy a few adjustable steel sawhorses at the hardware store, and lay a flat piece of ¼ to ⅜-inch-thick (6 to 9.5 mm) metal across them. Do not lay metal on top of a wooden table, however, as the wood could eventually become a fire hazard.

Pliers

Needle-nose pliers have narrow jaws that can easily reach inside the welding gun nozzle and pull out spatter that forms during the normal course of welding. These pliers also have a wire cutter that will allow you to snip off the end of the wire between welds when necessary. Tool lovers will be thrilled to know that you can purchase specialty welding pliers that come with a MIG tip remover in addition to various bells and whistles. However, for most purposes, needle-nose pliers work just fine.

From top: various needle-nose pliers and wrench

Wrench

As you work, you'll learn more about what you actually need. A wrench is a handy tool for connecting the gas hose, and I keep an adjustable model in my shop at all times.

Clockwise from top: bench vise, C-clamp, and clamping pliers

Bricks

For holding down pieces or propping them up, bricks are quite useful. I always keep at least three in the shop.

Bench Vise

A bench vise is a piece of equipment that has one fixed jaw and another parallel jaw which moves toward or away from the fixed jaw by means of a movable screw. The jaw will tightly hold your metal pieces while you bend, grind, and even weld on them. Bench vises come in many different sizes, but a small one is sufficient while you're just getting started.

Clamps

As you work on your welding projects, you'll likely find yourself wishing you had another hand, or maybe even two. Whether you're welding chair legs or attaching the holder for a plant stand, your clamps will become valued "shop assistants." And you've got options. Magnetic clamps are strong and will allow you to position and hold pieces in place while you weld. C-clamps and clamping pliers also come in handy for a variety of tasks. I also use small crafting magnets to brace lightweight objects such as ball bearings while I weld them.

Band Saw

Safer and easier to use than the cut-off wheel on your angle grinder, a horizontal band saw is great for cutting rod and flat bar. Small saws usually employ a gravity-fed blade that falls in an arc around a pivot point. Follow your machine's manual to determine which kind of blade you need for cutting steel. Always wear eye protection when using the band saw, as pieces of steel can fly off at any time.

Square Ruler

Often called a right angle ruler, this tool will prove valuable for determining and distinguishing angles when you're attaching a chair or table leg.

Drill Press

When used correctly, a drill press will allow you to safely cut holes in steel. This tool is useful for making holes for light to pass through for a lamp or nail holes for hanging projects. A small, inexpensive drill press is perfect for beginners.

Welding Jig

Welding jigs help you easily repeat a particular form, holding your parts in place while you weld or bend them. The jig can be as simple as a brick or something you make for a specific purpose. In this book, for example, we'll use a jig to hold the legs in place while we weld together a candleholder. It allows the legs, which are the base of the candleholder, to be braced and welded at the correct angle in order to build the rest of the candleholder on top.

Welding jig made from scrap metal

Materials

Before you get busy creating your masterpiece, you'll need to gather a few materials that all welding projects require. The most important material, of course, is metal, and here's some important information to know before you make any major purchases.

Metal

All metals can be classified as either ferrous or non-ferrous. Ferrous metal, such as mild steel (ordinary welding steel), contains iron and is magnetic, making it great for welding. To help me remember that ferrous metals are magnetic, I picture a steel ferret with magnets stuck all over it. Non-ferrous metals—aluminum, copper, brass, lead, and tin—do not contain iron and are not magnetic. Because MIG welding is a process of melting and joining metal, the pieces that

you weld need to be made up of the same ferrous material. For example, you can't weld aluminum, copper, brass, and steel all together mostly because each type of metal has different melting properties.

Readily available at big hardware stores and from architectural supply houses, mild steel is perfect for MIG welding. There are two types of mild steel: cold roll and hot roll. Hot roll steel is more common and less expensive, but, when dimensional precision is necessary, the cold roll process is more exacting. For our purposes, either one will work equally well, and, since they have the same composition, they weld easily to each other.

You'll need to check your manual before you bring home steel that might be too thick for the limits of your welding machine. I do, however, recommend

Bins of weldable steel at a hardware store

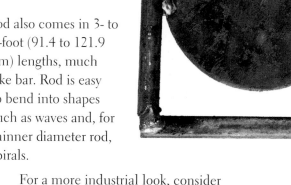

that you experiment for yourself before making your final decision about what your machine can and cannot handle; you might be pleasantly surprised. In the past, I've successfully welded together parts that, if I had gone completely by my manual, I would have never attempted to join.

Bar

Although there are many different sizes of steel bar stock, they are all either square or rectangular in shape. At the bigger hardware stores, bars usually come in 3- or 4-foot (91.4 or 121.9 cm) lengths and in a variety of widths and thicknesses such as 2 inches wide by ⅛ inch thick (5 cm by 3 mm) or 1 inch wide by ¼ inch thick (2.5 cm by 6 mm). I've used bar stock to make many of the projects in this book, and in each case I've indicated the thickness. I encourage you to experiment with other thicknesses, especially if the one I've recommended is not available.

Rod

Steel rod is round in shape and is available in many different diameters such as ⅛ inch (3 mm), ³⁄₁₆ inch (5 mm), and ¼ inch (6 mm). At large hardware stores,

rod also comes in 3- to 4-foot (91.4 to 121.9 cm) lengths, much like bar. Rod is easy to bend into shapes such as waves and, for thinner diameter rod, spirals.

For a more industrial look, consider using rebar in your piece. Originally called reinforcement bar for concrete, common rebar is made of unfinished weldable steel. The main thing to consider with rebar—identifiable by its distinct ridges—is whether or not your machine can handle the thickness.

Sheet Metal

Useful in all kinds of welding projects, flat sheet metal comes in a variety of thicknesses ranging from 30 to 8 gauge, with 30 being the thinnest. While you're learning, avoid sheet metal that is thinner than 18 gauge, because it's easy to burn holes through.

From left: various sizes of rod, bar, and rebar

Architectural Steel

Architectural supply houses provide materials primarily for the manufacturing of gates, fences, and railings. If there's no supply house close to where you live, you can easily find one online. Scrolls, finials, and ball bearings are among my personal favorites, and I use them in much of my work. When purchasing from these sources, pay close attention to the material composition. Frequently aluminum, brass, and cast iron parts are mixed in with steel parts.

Found Steel

Scrapyards are wonderful places to find parts for your projects, but it can be easy to become overwhelmed with all those mountains of metal. You might just want to look around the first time you go. The next visit, have a specific task or part in mind; you'll be amazed at the variety of materials you can find once you train your eye to see potential in the parts. Whenever you go, make sure you practice scrapyard safety (see page 24).

To help you sort things out, take a magnet with you so you can check to see what materials will work for welding. It can be hard to know what exactly you're looking at, but you definitely want to avoid some types of metal. For example, you might be fooled by stainless steel, which is magnetic; you can usually distinguish it by its suspicious lack of rust. Aluminum is lighter in weight than steel and will also be free of rust. Copper and brass can be spotted by their color. Galvanized metal has a whitish coating on top,

Assorted architectural steel parts, including ball bearings, spirals, finials, and leaves

and cast iron can have a porous quality to it. Sometimes even metal experts have a hard time telling if a piece is weldable; if you're unsure, just ask the people that work there before you make your purchase.

If a piece of weldable steel has paint or rust on it, you may still be able to use it; simply grind off a 1-inch (2.5 cm) spot for your weld. Take care when grinding painted metal as the paint could contain lead and be dangerous to breathe in; this is also true for galvanized steel, which has a zinc coating. A good dust mask and plenty of ventilation in your workspace will help ensure your safety while grinding and welding on scrapyard metal.

Found steel parts

Wire

Wire comes in a variety of sizes. You'll need to pick the appropriate size based on the thickness of the material you want to weld. As a general guide, .023-inch (0.6 mm) wire is used for welding thin material, and .035-inch (0.9 mm) wire is used for thicker material. For the projects in this book, .030-inch (0.8 mm) wire is your best bet.

Gas

The same oxygen that's great for breathing is not great when mixed with the weld area, due largely to impurities in the air. This is why a shielding gas is used. When you pull the trigger, gas is released and flows out to surround the spot where you are welding. The gas normally used for MIG welding is a mixture of argon and carbon dioxide. You can see for yourself what a weld looks like without the shielding gas by welding on a piece of scrap before you open up your gas tank.

Nozzle Dip

During welding, splatter can form and stick to the inside of the nozzle, resulting in buildup that can interfere with the welding process or get stuck on your project. By welding a little bit to warm up the welding gun and then sticking the gun into nozzle dip—a gel-like substance that melts and covers the gun—you can keep the splatter from sticking inside the nozzle and causing problems.

Chalk or Soapstone

You'll need to measure and mark your work when planning and making cuts. Chalk is readily available, and soapstone can be purchased at bigger hardware stores; either one works well for making non-permanent marks on steel.

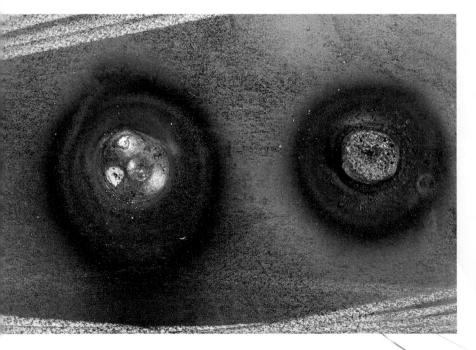

Tack welds made with shielding gas (left) and without (right)

Welding wire

Safety

It's important to have what I call a "warrior's mind" when you're in the shop. As you work, stay relaxed yet keenly aware of all the possible problems that can occur. Try to maintain a healthy respect for your tools and materials, some of which are hazardous. Because of this, there are possible dangers that you should be aware of: harmful fumes and smoke due to poor ventilation, flying debris, fire hazards, and the effects from ultraviolet and infrared light when welding without protection.

If you're working in an area with others nearby, consider purchasing a welding screen to protect them from debris and UV light.

The author in her safety gear

Fire Extinguisher

Sparks can and will fly, often farther than you think possible. Keep a fire extinguisher in an accessible location in your shop at all times. I always wondered if I would know how to use a fire extinguisher under fire, and I'm here to testify that they are extremely user friendly in the case of an emergency (!).

What To Wear

Although easy to avoid, the most common MIG welding shop injuries are burns—caused from handling the steel before it's had a chance to cool—and injuries caused by small bits of flying debris. You can avoid many of these types of injuries by wearing the appropriate gear.

Hearing Protection

A necessity when you're working with the angle grinder, there are a few different types of hearing protection, at a variety of prices, at most hardware stores. Plugs that have a band or cord to connect them keep your ear protection from easily getting lost or misplaced.

Hearing protection

Helmet

With a standard welding helmet, you have to flip your head to bring your helmet down when the flash occurs. Auto-darkening helmets—with a lens that darkens as you begin welding—are common these days, and good, reliable ones can be found fairly inexpensively online. In both types, the dark lens allows you to see what is going on without hurting your eyes, but it may take a while to get used to seeing through the darkness. Check for cracks in the shield of your helmet every once in awhile by holding it up to the light and looking through the shield.

Clear Protective Shield and Safety Glasses

Protective shields and safety glasses or goggles are a must when grinding, cutting, drilling, and for any tasks other than welding (when your eyes are protected by your helmet). Safety goggles completely enclose the eye in a protective cover with which many shop workers feel most comfortable. For the face and neck, I also like the added protection of a clear full-face shield. Remember to ensure the safety of people visiting your work area by providing them with safety gear or making sure they are a safe distance away.

Clear protective shield and dust mask

Dust Mask

A disposable dust mask, like the kind you can purchase in the paint department of your local hardware store, will protect you from breathing in the particles released from grinding. It's also important to make sure there is plenty of ventilation in the area where you are working.

Gloves

Gloves serve as a second skin for shop workers. In addition to protecting you from sharp objects, your gloves will protect you from the heat given off during the welding process. When just starting out, any leather gloves will do. However, welding gloves provide insulation to protect your fingers when you can't resist picking up those hot pieces, and the added wrist protection can prove invaluable when you accidentally rest your arm on hot metal. It might be difficult for women and small men to find gloves that fit properly. I had to special order my first pair through an industrial supply store, but now you can easily find small sizes online.

Proper Clothing

Protect yourself by wearing natural fibers such as leather, cotton, or wool. Polyester catches fire easily and will quickly melt onto whatever it comes in contact with.

The same is true for rubber shoes, so wear only leather shoes, preferably without laces, while working in the shop. Long sleeves will protect your bare skin from flying debris and also from the effects of ultraviolet and infrared rays from electric welders. Denim jeans are good, although watch out for frayed edges that could quickly catch fire from a stray spark.

Sun Block

Because of the ultraviolet light that is produced during the welding process, a day in the shop is like spending many hours in the sun. Sun block—with an SPF of at least 30—will help protect your face, neck, and hands from damage. You might also want to keep a bottle of Aloe Vera, or the plant itself, nearby for burn relief.

THE BASICS

Scrapyard Safety

Each scrapyard has its own set of rules and regulations regarding public hours, where you're allowed to wander, and the wearing of certain safety gear, so you might want to call in advance. Some scrapyards provide safety vests and hard hats, and they all require close-toed shoes. Gloves are also a good idea, since you'll likely be picking up rusty or dirty parts.

When you're out there digging around, don't be tempted to choose any closed cylinders and containers, no matter how perfect they might appear for a particular project. Regardless of what the container was used for or how long it's been empty, the potential for explosion due to flammable liquids and vapors is simply too high to make the risk worthwhile.

Getting Started

You've purchased the tools you need to get started, you're wearing clothing that is suitable, and you have some practice material nearby. Congratulations! In no time at all you will be creating those awe-inspiring pieces you have dreamt about. There are just a few more things to do before you get down to business.

Setting Up Your Shop

Ventilation is one of the most important considerations when setting up your workspace. With MIG welding, the ideal shop location is an area where the air is still and free from wind that can displace the shielding gas. The area will also need to be kept free from flammable materials such as rags, wood, and paper, as sparks can be unpredictable. For this reason also, concrete or dirt is the best kind of flooring. A heated garage is a good choice for a welding shop, while basements are not, due to the dangers of fire and compressed gases so close to your living space.

Getting to Know Your Welding Machine

The manual that comes with your welding machine will give you loads of valuable information in order to get started. Take time to read through the manual thoroughly so you are familiar with all the different parts of your machine. Allow yourself time to let everything sink in; it will, I promise. You can also refer to page 11 for a quick review on each part of the machine.

Inside your welding machine, there will be a chart that gives you information about the heat and wire speed settings

Other Welding Methods

Commonly referred to as **gas welding**, oxyacetylene welding is used for both joining and cutting. In this process, two pieces of metal are held close together, and the touching edges are melted and fused by the welder's flame with or without the addition of filler rod. Oxygen/acetylene welding can be difficult for beginners because it requires a two-handed welding technique and many safety rules for storing and using gases.

When high-quality, precision welds are needed—like those for airplane and racing team technicians—**TIG welding** is the best choice. The process, also called Heliarc, uses tungsten inert gas and requires a fair amount of coordination to manipulate the torch with one hand, the filler rod with the other, and, with many machines, a foot-operated amperage control.

Stick welding gets its name from the use of a coated electrode—or "stick"—that, when heated, releases a shielding gas and mixes with the metal of the two pieces you're joining to form the weld. Stick welding doesn't work well for metals that are thinner than 1/8 inch (3 mm), and it requires frequent rod changing and a fair amount of practice in both striking and holding the arc.

Flux-cored welding uses wire that contains both shielding gas and filler materials that, when burned by the heat of the arc, produces the weld. This method is great for working outdoors in unstable or windy conditions, although it can be messy. Flux-cored welding produces spatter and slag, and it can also easily burn through thinner materials.

specific to your machine. These settings are a good starting point, and over time you'll learn what works best for you and your projects. In general, the thicker the metal you are welding, the more heat will be required. Because of the higher temperature, the wire will have to come out of the gun faster to keep up with the speed of the melting and joining. This means that a higher wire speed setting is necessary for higher heat settings. Conversely, a lower wire speed setting is best for lower heat settings when welding on thinner materials.

Preparing Your Metal

The amount of preparation you'll need to do depends on where you got your materials. Parts purchased from stores or architectural supply companies come ready to weld as is, meaning you can start welding and creating with these pieces immediately. However, scrapyard treasures and other found parts can be another story altogether.

Most welding instructors will tell you that steel must be completely free from rust in order for it to bond properly. This is absolutely necessary when you're welding on something such as a bridge,

an industrial storage tank, or a trailer hitch. I have found that a small amount of rust mixed in with some non-rusted metal does not significantly impact the quality of welds for most small-scale creative welding projects. If you find that you do want to weld on a totally rusted piece, simply use the flap disc on your angle grinder to clean off a 1-inch (2.5 cm) spot where you will place your weld. You'll need to clean off another similar-sized area where you will attach your grounding clamp.

Pieces made with found steel components

Found steel cleaned
with angle grinder

Techniques

Once you've prepared your materials, you still have a little more work to do before you're ready to begin welding the pieces together. Many of these techniques are covered in the Two of a Kind project on page 42, which makes it a great starting project.

Cutting Metal

Unless you find the absolute perfect piece for your project, you'll probably need to cut pieces to size in order to create your finished work. Whether you're cutting rod or flat bar, you'll need to make sure you wear the proper protective gear—eye protection, ear protection, face shield, and gloves—and follow all the directions on your angle grinder carefully, including the use of the wheel guard. You can also use a band saw for this type of metal cutting; always follow the manufacturer's instructions and safety precautions.

This process works well for cutting both rod and wire, and the same steps, with a few exceptions, apply for cutting flat bar. However, for some wire, you might be able to use your needle-nose pliers if the gauge is thin enough.

1. Measure and place a chalk mark at the place you want to cut.

2. Place the rod or wire in your bench vise and tighten securely about 2 inches (5 cm) away from the cut mark.

3. Firmly grip the angle grinder, with the cut-off wheel attached, with both hands and hold it a comfortable distance away from your body. Hold the grinder about ½ inch (1.3 cm) above the chalk mark and press the on button. Gently bring the angle grinder down to the chalk mark and allow the wheel to do the work of cutting (photo a).

4. For flat bar, you will be starting at one end of the line and moving slowly across it. Don't forget to breathe and to have patience while your grinder is doing the work and your hands remain steady.

5. Stop to see if you have cut enough to be able to bend the metal back and forth and break it off at the cut line (photo b). If you discover that it is not ready to break off, loosen the bench vise and twist the rod or wire around until the uncut part is facing up, re-tighten the metal in the vise, and repeat steps 3 and 4. For flat bar, you'll simply go over your cut line again.

Bending Rod

Because steel is flexible, it can be bent into various shapes. When learning how to bend rod, practice really does makes perfect. It takes time to understand exactly where the bend will occur in relation to where the rod is being clamped or braced. Bending depends on leverage. No matter how many reps or yoga poses you do at the gym, the human hand and arm can only do so much. There are a couple of techniques you can use to increase your leverage.

1. To secure the rod before you bend it, clamp it in a bench vise (photo c) or drill a ½- to 1-inch (1.3 to 2.5 cm) hole into your worktable and feed your rod through it to provide a strong grip for bending.

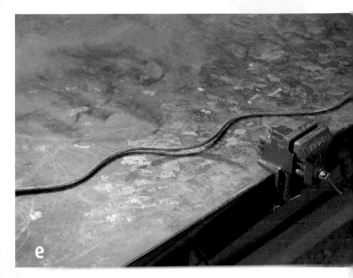

2. Depending on the desired result, start at one end by bending the metal slightly, either with your hands (photo d) or using a bending tube (see sidebar).

3. Unclamp the rod, move the rod ½ inch (1.3 cm) up in the vise, and re-clamp it in place for your next bend. Continue progressing up the rod in ½-inch (1.3 cm) increments (photo e). For smaller bends, progress at ¼-inch (6 mm) increments. It can be tricky to undo a bend, so stop and examine your results as you work.

Using a Bending Tube

If you are having trouble bending thick rod with your hands, use a bending tube. A bending tube is simply a piece of ¾- to 1-inch (1.9 to 2.5 cm) steel pipe that you cut with your angle grinder or band saw to different lengths; you might even be able to find one lying around the scrapyard. The longer the pipe, the more leverage you'll have and the thicker rod you can bend.

Bracing Your Pieces

Before you start welding anything, you'll want to make sure your pieces are positioned correctly. Different welders have different methods. Some use magnets, bricks (photo f), or clamps to hold their pieces in place. This is probably the best approach if you need perfect angles for your finished piece to function properly. I prefer to hold my pieces with my free hand in most instances. When a more exact angle is required I often use a square ruler.

Welding

There are many different kinds of joints and welds when welding for industrial purposes, but understanding them is not necessary when welding for fun and creativity. The main types of welds we'll be concerned with are what I'm calling short and long tack welds. Then we'll cover how to run a bead and how to weld at a corner joint.

Short Tack Weld

A short tack weld is used largely for temporary positioning and is basically the same as a long tack weld, just a little smaller. Short tack welds are better than long tack welds when you want to make sure the weld can be easily broken apart if it's not exactly where you want it to be and when you don't want to create too much heat that could move or warp your material. Before you get started, make sure you are dressed properly (see page 23). Then gather your helmet and gloves, and grab a practice piece of steel. The 3 x ¼-inch (7.6 cm x 6 mm) flat bar used in Two of a Kind (page 42) is a good thickness for practice.

1. Unscrew the valve on your gas tank, and set your heat and wire speed.

2. Brace your pieces in place (photo g), attach the grounding clamp to your worktable or your workpiece, and put on your helmet. Prior to starting your weld, snip off the end of the wire, called "stickout," if it sticks out the end of the gun farther than about ¼ inch (6 mm) (photo h). The ideal stickout is about ⅛ to ¼ inch (3 to 6 mm). If the wire sticks out too long, you won't be able to get in close enough to your workpiece to properly start your weld.

3. Hold the gun at a 45° angle about ¼ inch (6 mm) away from the workpiece (photo i).

4. With your welding gun held steady, lower your helmet (if you have an auto darkening helmet, skip this step) and firmly press the trigger. It might take a while to get used to holding your welding gun at the exact spot where you want to place your weld when everything goes dark. With a little practice, this will become second nature.

g

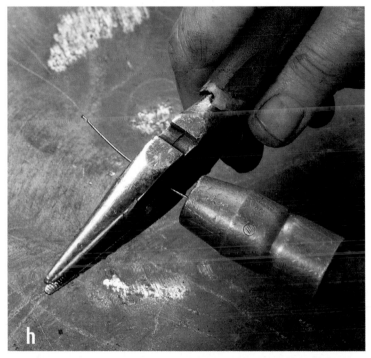

h

TIP

Don't be afraid to get your head (with your helmet down, of course) close to where you'll be welding. About 16 to 18 inches (40.6 to 45.7 cm) or closer is a good distance in order to be able to see what's going on. When you get more comfortable with the process, you will begin to see the base metal melting (called the "puddle") while you are adding the metal from the wire feed. You will know you're making a good weld when it sounds like eggs frying.

i

5. Hold the trigger for one full second for a short tack weld (photo j).

6. Release the trigger and inspect your first weld (photo k). It should look smooth and flat.

Long Tack Weld

Long tack welds are used when you're ready to place your final weld. Normally you'll be going over your short tack welds with longer tack welds to strengthen them, as well as adding additional long tack welds to your piece as necessary. The steps for this type of weld are the same as those for the short tack weld, except that you'll weld for a full two seconds without moving the gun (photo l).

Troubleshooting

Are you still running into issues after you've experimented with these techniques? Don't worry. Here are some common problems and how to fix them.

Problem	Try This
You follow the recommended heat and wire settings, but the wire spatters back, not wanting to melt and join.	If there's a lot of rust, your piece may be too corroded for the weld to join. Carefully grind the spot you're welding down to clean metal. Check to see that the welding clamp is connected on clean metal that's close to the piece you're welding. Turn down the wire setting a little to give the wire a better chance to melt before it connects with the piece. Turn up the heating setting slightly.
Your wire gets stuck on the piece after welding—an issue called stubbing.	Try releasing the wire by gently moving the welding gun side to side until it snaps off, or cut it off with pliers. Turn down the wire setting a little. Turn up the heat setting slightly. Make sure to let go of the trigger immediately after the weld is made.
Your wire melts onto the end of the contact tip and gets stuck.	Do not continue depressing the trigger. Turn off your machine and remove the nozzle from the welding gun. Gently unscrew the contact tip, keeping hold of it the entire time. Once unscrewed, turn on your machine and press the trigger to advance the wire a few inches. Turn off the machine again, cut off the tip, and screw in a new contact tip. Replace the nozzle. Turn down your heat settings a little and try again. Hold the gun a little farther from your work; try ¼ inch (6 mm). If it happens again, turn up your wire speed a little.

Running a Decorative Bead

For the projects in this book, running a bead is used largely for decorative purposes on the surface of your metal. In this process, keep continuous pressure on the trigger for the entire length of the weld. Practice first on some flat bar to get the hang of things.

1. Unscrew the valve on your gas tank, and set your heat and wire speed.

2. Attach the grounding clamp to your worktable or your workpiece, and put on your helmet. Snip off the end of the stickout.

3. Lay the pieces out on your worktable, and secure them with a brick or clamp.

4. Depending on the type of design you want to create, hold the welding gun at a 45° angle and move it slowly over the surface of the steel to draw designs (photo m). Start out by making 1- or 2-inch (2.5 or 5 cm) lines, slowly weaving the tip of the welding gun side to side in a zigzag pattern to create a line as you work across the metal, and then experiment until you get your desired effect (photo n).

5. Grind the surface of the welds using the flap disc on your angle grinder (photo o).

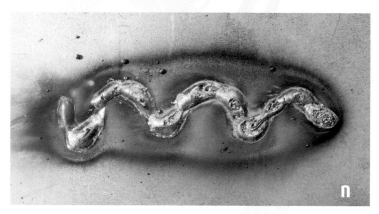

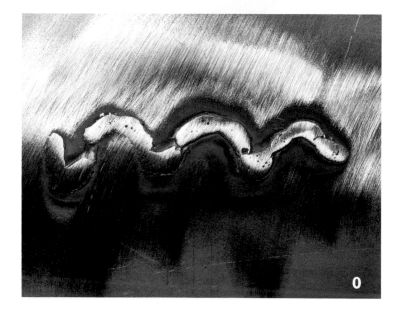

Welding a Corner Joint

To make many of the projects in this book, you'll need to weld two pieces of metal together at a corner joint.

1. Once you've cut and prepared your pieces, lay them out on the worktable exactly as you want them to be after they are welded together. If there is a front side and back side to the pieces, lay them facedown so you're welding on the back side. Clamp the pieces down or use bricks to hold them in place.

2. Refer to the diagram on your machine for heat and wire speed settings, then set your machine and put on your helmet.

3. Set your grounding clamp, trim the stickout, and aim the welding gun at a point along where the two pieces meet. Place one short tack weld in the center of each area where the sections are touching.

4. Go back and add an additional two or three short tack welds to each section to strengthen the piece, and then go over the short tack welds with long tack welds (photo p). Flip the piece right side up.

Finishing

There are a number of ways to decorate and finish your projects. A short trip to the hardware or craft store can give you all kinds of finishing ideas, such as metal paints and spray-on textured paints. Steel will rust if left outdoors, producing a very attractive natural patina. You can also create designs and texture on the surface of your work with the tools you already have in your shop.

Paints and Patinas

If you follow the directions carefully, paint and patina finishes can last a year or more outside before eventually being overtaken by rust. Have fun experimenting; there are countless products, such as hammered texture spray paint (see right), for creating various patinas on your metal pieces.

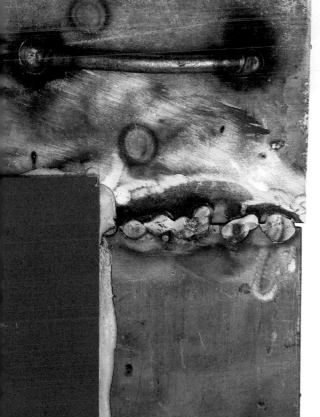

p

Rust

I love the natural texture that rust creates on mild steel. I feel it gives a piece character and makes it more interesting to look at. There are many ways to create a rust patina. The easiest method is to simply clean the steel with soap and water—to make sure all the oil is off—and then spray it with water or a mixture of one part water, one part hydrogen peroxide, every few hours or so. Within a day or so you will begin to see a beautiful rust patina forming.

Spatter Texturing

Your MIG welder can also be used to decorate your projects. Spatter texturing lightly deposits weld material on the surface of your piece, creating an interesting textured look.

For spatter texturing, you'll need to turn your heat setting down and your wire speed setting up slightly higher than recommended on your machine. Hold your welding gun about ½ inch (1.3 cm) from your piece, and, while depressing the trigger, quickly move the gun side to side over the surface. Have fun experimenting until you achieve the look you want. It's a great way to learn, and sometimes those creative "accidents" produce the perfect look for your piece.

Luster, Brush, and Etching

You can achieve various finishes by using different wheels on your angle grinder. The wire wheel brush is great for getting rid of bits of slag left over from welding, and it's able to reach into areas that the flatter discs can't get to. Sometimes I simply prefer the luster that the wire wheel gives to steel, which is less invasive than the discs.

Hard grinding discs cut deep into the surface of the steel. Discs are great for etching designs and for grinding steel when it's not appropriate to use the cut off wheel or band saw. Flap discs are my favorite for smoothing and creating a silvery brushed finish.

Protective Coatings

No matter what you do, steel is going to eventually rust if you leave it outside for any length of time. Serious metal workers that create work for outside use—such as railings and gates—go to great lengths to keep the rust from taking over. Sandblasting, powder coating, and galvanizing are a few strategies often used for keeping rust at bay.

If you're someone who absolutely cannot stand rust but you don't want to put in the effort to avoid it, keep your projects indoors. Even inside, you'll want to give your work a protective clear coating, as the humidity in the air will begin to rust unprotected steel. Do this by spraying your creation with clear polyurethane.

Some metalworkers swear by all-purpose lubricant; rubbed on with a soft cloth, it can protect steel from rust. Boiled linseed oil is also a frequent choice and can be found at craft stores. If you want to get fancy, a three-part mixture of equal amounts of mineral spirits, Japan dryer, and boiled linseed oil is a time-tested recipe for avoiding rust.

Welding for Creativity

Welding is a great way to tap into your own artistic abilities. There are a few things you can do to help stimulate your creative process, especially when you're trying to create your own designs. For example, let's use a picture frame.

First, capture your ideas. You will soon notice that picture frames are everywhere, so carry a notepad with you wherever you go. Store a pencil and paper near your bedside for those hard-to-catch ideas that surface upon awakening. You might also try keeping a simple voice-recording device with you in the car; it's much safer than trying to draw while driving!

Second, prime the pump. After you've completed the picture frame project in this book and you're ready to make your own unique design, start by collecting as many examples of picture frames as you can. Pore through magazines, and look at examples online. Go to stores, and make notes about the different types of frames that you see. Take photographs if you can.

Third, take time to study. Notice what parts go into making a picture frame. If you see one you like, don't just copy the piece exactly; study it to determine what makes it so appealing to you. Is it the finish that you like so much or the scrolls along the bottom? Look at how the various functions are achieved—such as the stand or the wall hanger—as well as how designs are applied.

Fourth, be willing to take risks. Don't worry if it seems foolish to try to get your picture frame to stand out from the wall a foot or to combine the lamp and picture frame in one piece. You might even see if you can incorporate photos into a three-panel screen project, or have your favorite photos printed onto the fabric that you use to make a bench seat.

And finally, remember to give yourself plenty of support and encouragement. Be kind to your creative self. You might even print this quote and hang it somewhere near your workspace as a reminder: "I give myself permission to experiment, take risks, play, explore, invent, make mistakes, and have fun!"

Easy
Projects

A pair of freeform bookends is just the thing for holding all your favorite books. Creating the abstract design is easy: simply use the welding gun like a paintbrush.

Two of a Kind

Materials

18 inches (45.7 cm) of 3 x ¼ inch (7.6 x 0.6 cm) flat bar

Tools & Supplies

MIG Welder Kit (page 11)
Angle Grinder Kit (page 12)
Clear acrylic spray

Step by Step

1. Measure, chalk, and cut the flat bar to make two 4-inch (10.2 cm) pieces and two 5-inch (12.7 cm) pieces. Grind the ends of each piece smooth with the angle grinder.

2. Secure one of the 4-inch (10.2 cm) pieces to your work surface with a C-clamp. Attach the flap disc to the angle grinder, and go over the entire surface using a small circular motion to soften the sharp edges and give the steel a brushed finish. Repeat this process for the other piece and for both sides of each 5-inch (12.7 cm) piece.

3. Refer to the diagram on your machine for heat and wire speed settings, and then set your machine.

4. Lay one of your 5-inch (12.7 cm) pieces out on the welding table. Make a few small spot welds on the surface, and then experiment by moving your welding gun slowly over the surface to create designs, not worrying so much how it looks. Secure the piece with a C-clamp, and grind the surface using the flap disc. Repeat this process with the other 5-inch (12.7 cm) piece.

5. Using a brick, prop one of the 5-inch (12.7 cm) pieces on the table at a 90° angle. Lay one of your 4-inch (10.2 cm) pieces flat on the table so it butts up against the longer piece.

6. Working on the inside of the joint, make a short tack weld to set your piece. If it's not where you want it, you can easily break the weld by hitting it hard on the worktable or by grinding off a small amount of the tack weld and then breaking it apart.

7. Finish welding the joint with two or three long tack welds (photo a). Repeat steps 5 through 7 with the other two pieces. Go over all the surfaces again using the angle grinder and a soft flap disc. Coat the piece with clear acrylic spray to guard against rusting.

Picture This

A special photo should have a special frame—this one even has its own stand. Spot welds and bent rod are employed to create a one-of-a-kind finish.

Materials

22 inches (55.9 cm) of 2 x ¼-inch
(5 cm x 6 mm) flat bar
14 inches (35.6 cm) of ³⁄₁₆-inch
(5 mm) wire

Tools & Supplies

MIG Welder Kit (page 11)
Angle Grinder Kit (page 12)
Clear acrylic spray
Small magnets

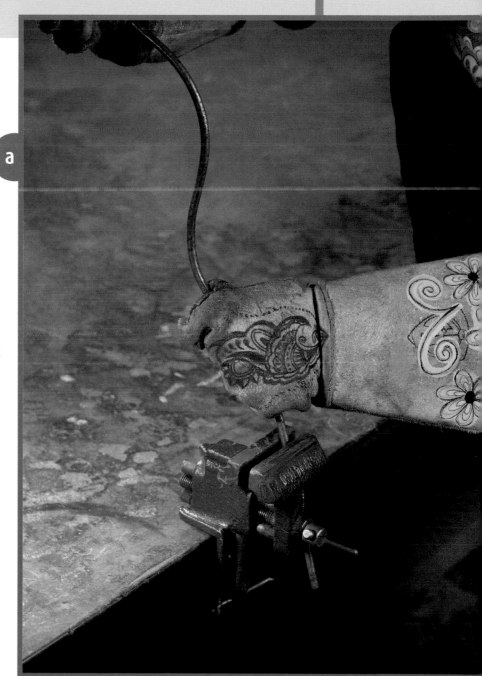

Step by Step

1. Measure, chalk, and cut the flat bar into the following measurements: two 7-inch (17.8 cm) sections, two 3-inch (7.6 cm) sections, and one 2-inch (5 cm) section.

2. Insert the wire into a bench vise and bend it slightly. Unclamp the wire, turn it around, re-clamp it, and bend it slightly in the opposite direction. Repeat this process a third time to create a zigzag pattern (photo a).

3. Clamp the bent wire into the bench vise, and measure and cut a 2-inch (5 cm) section from the end. Repeat this step until you have seven 2-inch (5 cm) pieces of wire.

4. Lay out the cut flat bar pieces facedown on your work surface. The 7-inch (17.8 cm) pieces serve as the frame's top and bottom, and the 3-inch (7.6 cm) pieces form the sides. Make sure all the pieces are touching. Refer to the diagram on your machine for heat and wire speed settings, and then set your machine.

b

5. Connect the pieces following the steps for making a corner weld (page 37).

6. Once you've welded the four pieces together, turn the frame faceup, and clamp it down on the edge of the worktable. Use the flap disc to go over the entire surface of the frame, taking care to soften the outside edges. To do this, you'll have to work your way around the frame, clamping and unclamping it in order to reach all the sides.

7. Use the welding gun to create small dots on the frame's surface. Don't worry that it doesn't look great yet; the surface will look quite different after you grind it with your flap disc in step 9.

8. Turn the frame over, so the backside is facing up again, and lay the zigzag wire pieces down along the top edge of the frame, overlapping the edge by ¼-inch (6 mm). Brace the zigzag pieces with a brick, and make short tack welds to secure each piece to the backside of the frame.

9. Turn the frame faceup, clamp it in place on your worktable, and use the flap disc to go over the dots and heat marks until you get the look you want.

10. To create the stand for the frame, hold it up on the work surface with the top tilted slightly back. Slide the small 2-inch (5 cm) square piece up against the bottom back of the frame, clamp both pieces onto the work surface, and weld them together (photo b).

11. Coat the frame with clear acrylic spray. Secure your photo on the backside of the frame using small magnets.

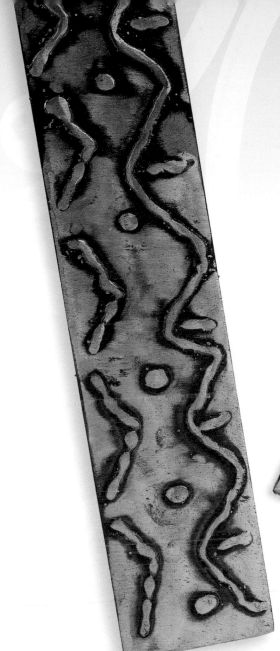

With just a few simple steps, you'll have a handy message board and matching magnets. This project is great for beginners and employs a neat trick to create the wavy designs.

Magnet Board

Materials

14 inches (35.6 cm) of 3 x ¼-inch (7.6 cm x 6 mm) flat steel bar [A]

5 inches (12.7 cm) of ⅛-inch (3 mm) rod

4 inches (10.2 cm) of 1 x ¼-inch (2.5 cm x 6 mm) flat steel bar [B]

Tools & Supplies

MIG Welder Kit (page 11)
Angle Grinder Kit (page 12)
Glue
Small magnets
Clear acrylic spray

Magnet Board

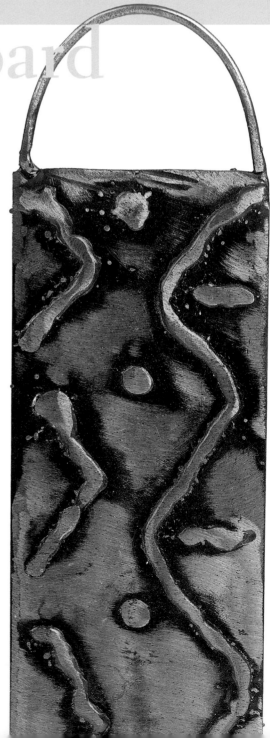

Step by Step

1. Measure, chalk, and cut a 14-inch (35.6 cm) piece of the flat steel bar A. Use the grinding disc on your angle grinder to smooth the cut edge.

Making the Board

2. Secure the piece to your work surface with a C-clamp. Working with a small circular motion, use a flap disc on your angle grinder to go over the entire metal surface to give it a brushed finish.

3. Refer to the diagram on your machine for heat and wire speed settings, and then set your machine.

4. Put on your helmet and make six dots spaced evenly down the center of the bar. Next, experiment by moving your welding gun slowly over the surface to create wavy designs, and then grind the surface with the flap disc.

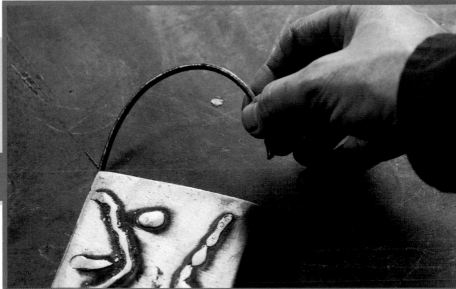

Creating the Hanger

5. Secure the rod in a bench vise, measure and chalk a 5-inch (12.7 cm) piece, and then use the cut-off wheel on your angle grinder to cut the piece at the chalk mark.

6. Insert one end of the piece into the bench vise, and bend it slightly following the steps on page 30. Continue bending the rod at ¼-inch (6 mm) intervals until it becomes a "U" shape and the ends are about 3 inches (7.6 cm) apart.

7. Lay the flat bar on the work surface, face side down, and brace the ends of the bent U-shape rod against the top end of the flat bar (photo a). Point the welding gun at the place where the rod meets the flat bar, and place a short tack weld. Repeat for the other side. Go over your tack welds with another short tack weld if it seems to need it in order to hold together.

Constructing the Magnets

8. Secure flat bar B in a bench vise, and place angled chalk marks at 1-inch (2.5 cm) intervals to create four odd-shaped square shapes. Use your angle grinder to cut the pieces at the chalk marks.

9. Decorate the square pieces with welded dots and lines (photo b). Next, place each one in the bench vise and grind the surface with the flap disc.

10. Glue magnets to the backs of the squares.

11. Spray the magnet board and the magnets with clear acrylic spray to guard against rust.

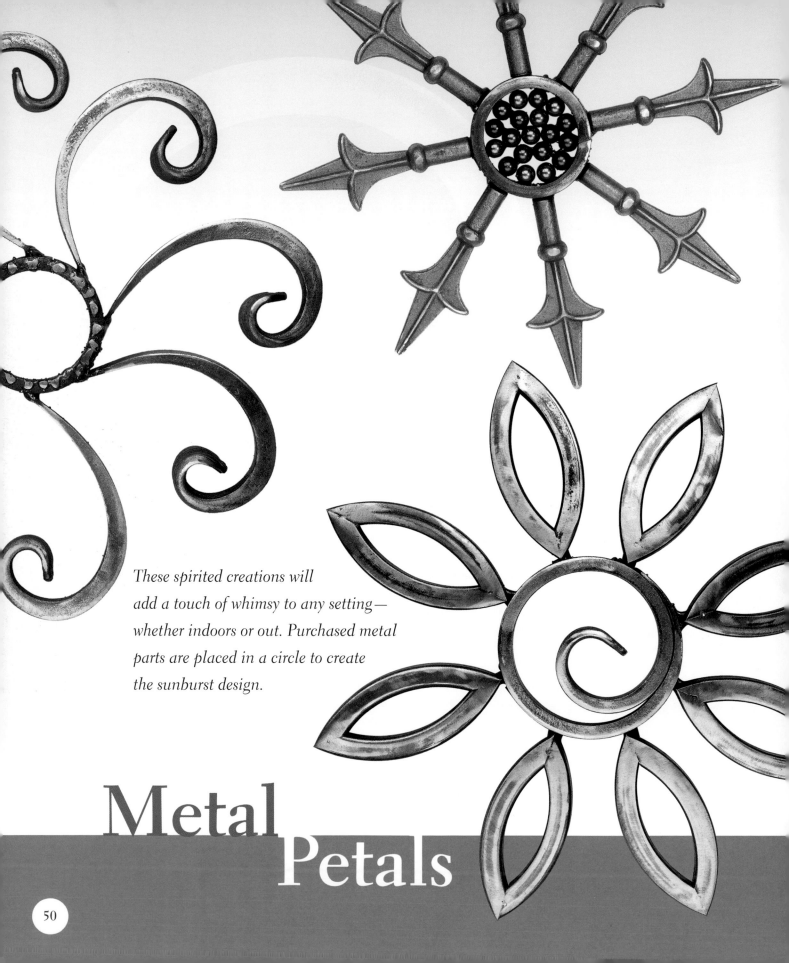

These spirited creations will add a touch of whimsy to any setting— whether indoors or out. Purchased metal parts are placed in a circle to create the sunburst design.

Metal
Petals

Materials

Metal ring or spiral, 3 to 5 inches (7.6 to 12.7 cm) in diameter
Metal rays, 4 to 8 inches (10.2 to 20.3 cm) in length

Tools & Supplies

MIG Welder Kit (page 11)
Angle Grinder Kit (page 12)
Clear acrylic spray

Step by Step

1. Lay out the metal ring (or spiral) on your worktable with the back facing up. Think of a clock, and place chalk marks where the 12, 3, 6, and 9 numbers would be. Then place chalk marks in between each mark you just made. You should now have eight marks that are an equal distance apart on the ring.

2. Lay out the ray shapes facedown with each one touching a chalk mark and radiating directly away from the center.

> **TIP**
>
> If your ray shapes and ring are different thicknesses, use angle grinder discs or pieces of scrap metal to prop the pieces up until they're all on the same level.

3. Brace the rays with bricks, and short tack weld where they touch the metal ring. Work your way around the ring until you've welded all the rays (photo a).

4. Turn the piece over and gently go over the front surface with the flap disc.

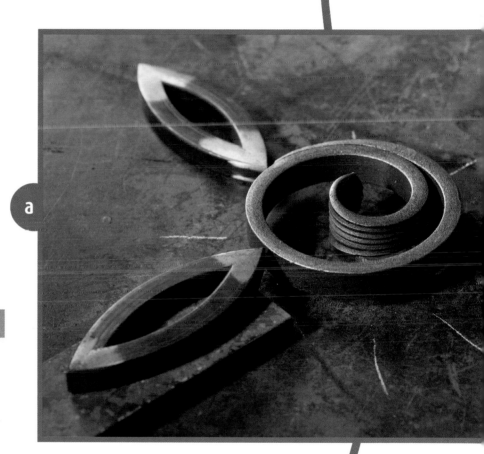

a

5. For a decorative touch, add deposits of weld to the perimeter of the ring, and then soften these with a flap disc (photo b).

6. Spray the flower with clear acrylic to prevent rust.

Separately or grouped, these stakes will add interest to your flower bed or container planting. Let them weather naturally for a pleasing aged patina.

Garden Stakes

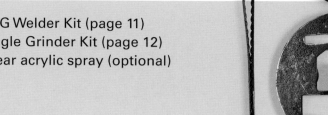

Materials

9 feet (274.3 cm) of ¼- to ½-inch
(6 mm to 1.3 cm) rod

5 to 10 decorative found metal
shapes, each under 1 foot (30.5
cm) in diameter

Tools & Supplies

MIG Welder Kit (page 11)
Angle Grinder Kit (page 12)
Clear acrylic spray (optional)

Garden
Stakes

Step by Step

1. Measure, chalk, and cut your rod until
you have three 3-foot (91.4 cm) pieces.

TIP

If you need to weld pieces together to make a 3-foot (91.4 cm) piece,
grind off any rust so you've got a clean spot to place your weld,
brace the pieces with bricks or clamps, and place a short tack
weld at the place where the two pieces of rod meet. Check that the
pieces are connected properly, and then go back over each short
tack weld with a long tack weld. Turn the rod over and place another
long tack weld on the opposite side of the rod. For added strength,
you can also place two long tack welds on the sides.

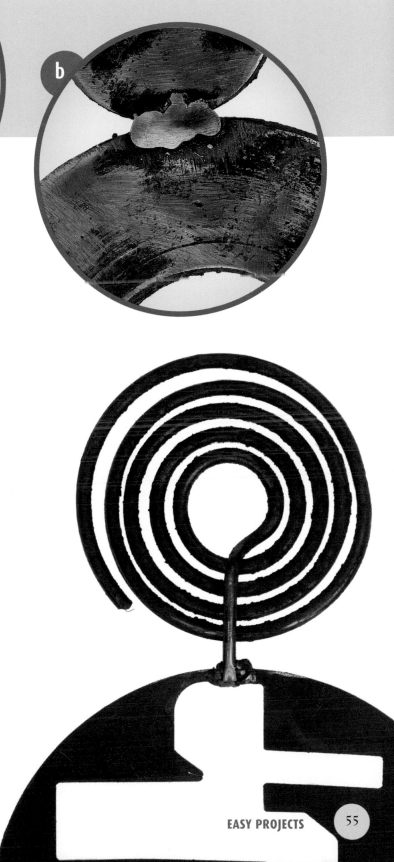

2. Secure your decorative shapes in the bench vise, and go over each piece with the flap disc or wire wheel brush to clean and finish them. If the pieces are particularly rusty, you will need to grind off a clean spot so you'll have a place to put your weld.

3. Refer to the diagram on your machine for heat and wire speed settings, and then set your machine.

4. Lay the decorative piece upside down on your worktable, and place the end of a rod on top, near the bottom of the piece. Weld the rod to the back of the piece (photo a).

5. Continue to weld the pieces onto the rod, adding pieces above the first one (photo b), to create the decorative design. Repeat the same steps for the other two garden stakes.

6. Finish the stakes in your chosen method, although leaving your stakes to gain their own natural rust patina is a fine choice.

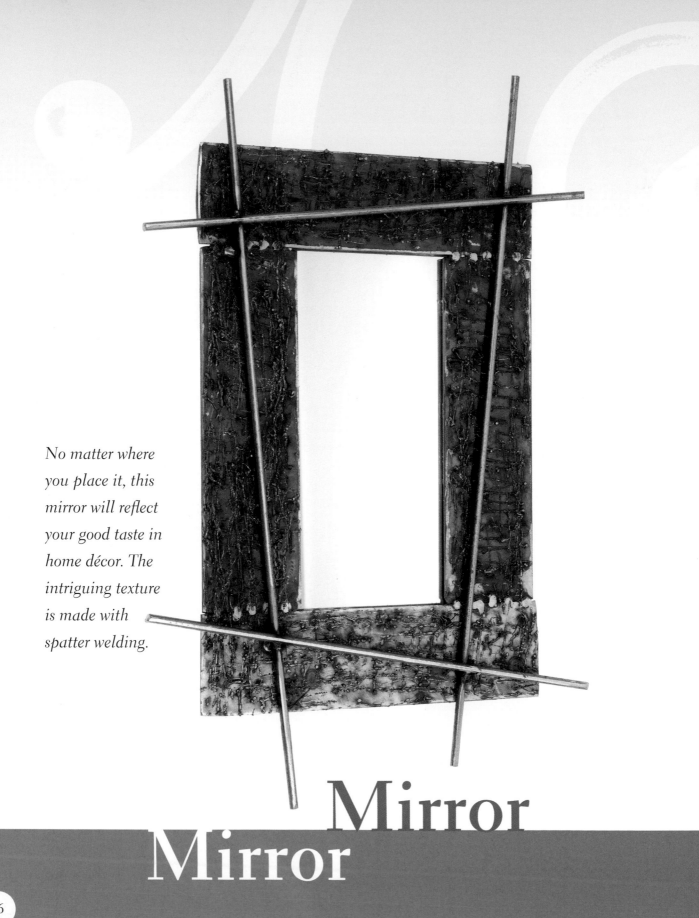

No matter where you place it, this mirror will reflect your good taste in home décor. The intriguing texture is made with spatter welding.

Mirror
Mirror

Materials

44 inches (111.8 cm) of 3 x ¾-inch (7.6 x 1.9 cm) flat bar
79 inches (200.7 cm) of ¼-inch (6 mm) rod
6 x 12-inch (15.2 x 30.5 cm) mirror, ¼ inch (6 mm) thick

Tools & Supplies

MIG Welder Kit (page 11)
Angle Grinder Kit (page 12)
Clear acrylic spray
Glue

Step by Step

1. Measure, chalk, and cut four 11-inch (27.9 cm) pieces from the flat bar. Secure one piece in the bench vise, and use the grinding disc to smooth the cut edges. Repeat for the other three pieces and set them all aside.

2. Working with the rod, measure, chalk, and cut two 22-inch (55.9 cm) pieces, two 14-inch (35.6 cm) pieces, one 3-inch (7.6 cm) piece, and two 2-inch (5 cm) pieces.

Creating the Frame

3. Lay the four bar pieces on the worktable with the pieces butted up against one another, using the project photo as a guide.

4. Refer to the machine's diagram for heat and wire speed settings, and then set your machine.

5. Put on your helmet and connect the pieces following the steps for making a corner weld on page 37 (photo a).

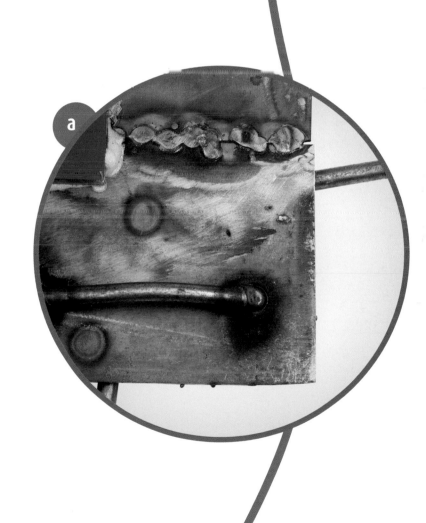

6. Clamp the frame facedown on your work surface, and pick up the mirror. Look to see where the mirror will be glued, and gently grind down the welds at the corners so the mirror can lie as flat as possible against the back of the frame. Turn the frame over, so the right side is up, and clamp it down securely. Go over the entire surface with the flap disc to achieve a brushed effect and to soften the edges.

7. Use the spatter welding technique (page 38) to create texture on the frame (photo b). Go over the surface of the frame gently with the flap disc to remove any wire that is not attached securely.

8. Using the photo as a guide, weld the two 22-inch (55.9 cm) rod sections directly to the frame. Next, lay the two 14-inch (35.6 cm) sections directly on top of the longer sections, and weld where they touch (photo c). Use the flap disc to soften the ends of the rod.

Making the Hangers

9. Secure the end of the 3-inch (7.6 cm) piece of rod in your bench vise, and tap it with a hammer to get a gentle curved shape. Follow the same steps for the two 2-inch (5 cm) pieces.

10. Turn the frame facedown, and lay the 3-inch (7.6 cm) piece in the center of the top of the frame, making sure it doesn't overlap where the mirror will be placed. Use a brick to hold the piece in place, and weld both sides of the hanger to the frame. Repeat with both of your remaining 2-inch (5 cm) pieces, placing them on both sides of the bottom of the frame (photo d). These pieces will make sure the mirror hangs a uniform distance off the wall.

11. Spray the frame with clear acrylic spray to guard against rusting. It's important that you do this before you attach the mirror, since acrylic spray will ruin the reflective surface. Glue the mirror to the back of the frame.

Fish Hooks

How much time do you spend everyday looking for those pesky keys? You'll always
know right where they are when you hang them from fish hooks made of rod.

Materials

48 inches (121.9 cm) of 3/16-inch (5 mm) rod [A]

1 inch (2.5 cm) of ½-inch (1.3 cm) round tubing

8 inches (20.3 cm) of ⁵⁄₁₆-inch (8 mm) rod [B]

24 inches (61 cm) of ⅛-inch (3 mm) rod [C]

12 inches (30.5 cm) of 1½ x ⅛-inch (3.8 cm x 3 mm) flat bar

Tools & Supplies

MIG Welder Kit (page 11)
Angle Grinder Kit (page 12)
Drill with ⅛-inch (3 mm) drill bit
Clear acrylic spray

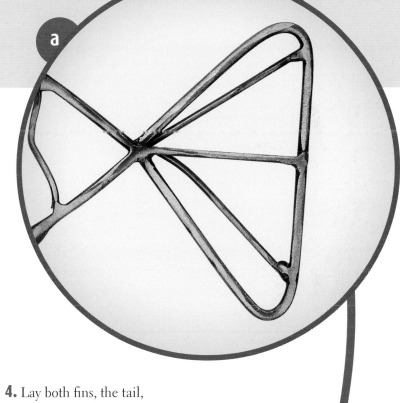

Step by Step

1. Measure, chalk, and cut all your pieces to the following measurements. With rod A, cut three 12-inch (30.5 cm) pieces, one 5-inch (12.7 cm) piece, and one 7-inch (17.8 cm) piece. Cut one ¾-inch-long (1.9 cm) piece from the tubing, and four 2-inch-long (5 cm) pieces of rod B.

2. To create the fish tail, mark a 12-inch (30.5 cm) piece of rod A at three spots: 3½, 8½, and 12 inches (8.9, 21.6, and 30.5 cm). Place it in the bench vise and bend at each mark to create a triangle shape. Place the other two 12-inch (30.5 cm) pieces in the bench vise and gently bend them to create the curves for the upper and lower portions of the fish body.

3. Bend the 7-inch (17.8 cm) piece of rod A to create the top fish fin, and bend the 5-inch (12.7 cm) piece to make the bottom fish fin.

4. Lay both fins, the tail, and the two fish body pieces out on your work surface, and weld the pieces together where they touch.

5. To fill in the tail section, measure the distance inside the tail and cut out three pieces from rod C. Weld the pieces in place, using the photo as a guide (photo a).

6. To create the wavy interior pieces for the fish, place rod C in the bench vise, bend it slightly, and unclamp it. Turn

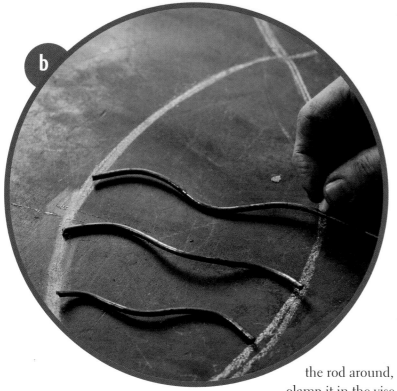

9. Finish welding the interior rods on the fish where the pieces touch. Clamp the fish down, and softly go over the whole shape with a flap disc to give it a finished look.

10. Lay the flat bar and the fish face down on the table so the bottom fin touches the top of the bar. Weld the pieces together where they touch.

11. Hold a 2-inch (5 cm) piece of rod B at an angle on the flat bar, and short tack weld. Attach the other three 2-inch (5 cm) pieces in the same fashion. When you're satisfied that you have them at the correct angle, go over all the welds to strengthen them (photo c).

the rod around, clamp it in the vise again, and bend it slightly in the opposite direction. Repeat the step again until the whole length of the rod is wavy.

12. Finish the piece as desired with a soft flap disc on your grinder. Drill holes in both ends of the flat bar to allow for wall mounting. Spray the piece with clear acrylic.

7. Lay the wavy rod across the fish and then, working from the top of the fish to the bottom, mark how much is needed to cover the distance, measuring each piece carefully as you move from left to right across the fish (photo b). Cut the wavy rod at each chalk mark.

8. Clamp the 12-inch (30.5 cm) piece of flat bar in a C-clamp. Next, use a flap disc to soften the edges, and go over the entire surface with a circular motion. Lay the piece on the welding table and apply spot welds for decoration, using the flap disc to finish the welds when you're done.

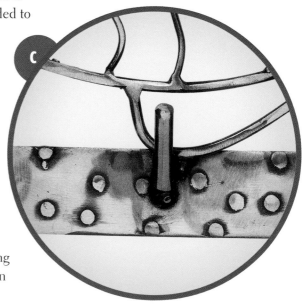

Intermediate Projects

Want a unique way to separate spaces in your home? Create a hinged partition from found rod and geometric objects.

Private Screening

Materials

70 feet (21.2 m) of ⅜-inch (9.5 mm) found rod
7 found geometric metal shapes
Four ½-inch (1.3 cm) nuts
12 inches (30.5 cm) of ¼-inch (6 mm) rod

Tools & Supplies

MIG Welder Kit (page 11)
Angle Grinder Kit (page 12)
Clear acrylic spray or oil
Small felt pads

Step by Step

1. Measure, chalk, and cut the found rod to the following dimensions: six 65-inch (165.1 cm) sections and six 14-inch (35.6 cm) sections. If your pieces need cleaning, grind the areas where you will be welding the rods together.

Creating the Panels

2. Working on one panel at a time, lay out two long rods (for the sides) and two short rods (for the top and bottom) on your worktable. Place the bottom rod 12 inches (30.5 cm) from the end of the long rods. Refer to the diagram on your machine for heat and wire speed settings, and then set your machine. Weld the pieces together at each corner where they touch. Repeat this step until you've created three panels.

3. Prepare all the found geometric shapes by grinding ½-inch (1.3 cm) spots where you'll place your weld. Use the same steps to clean and finish all the parts that you intend to add to the center of the screen panels.

Making the Hinges

4. Measure and chalk a spot 6 inches (15.2 cm) down from the top on the side edge of your center panel. Stand a nut upright (on one of its edges) on the worktable at the chalk mark so the nut is perpendicular to the panel's edge. Prop up your frame so the center of the nut and the rod are level. Weld the nut to the edge of the frame (photo a). Move down the rod, measure 12 inches (30.5 cm) up from the bottom, and repeat this step. Place nuts in the same spots on the other side of the panel.

5. Measure and cut four 1-inch (2.5 cm) and four 2-inch (5 cm) pieces of the ¼-inch (6 mm) rod. Weld one 1-inch (2.5 cm) piece and one 2-inch (5 cm) piece together to create an "L" shape. Repeat this step until you've created four "L" shapes.

6. Working on the inside edge of the left frame, measure 11½ inches (29.2 cm) down from the top, and weld the shorter end of the "L" shape to the outside of the frame. The 2-inch (5 cm) piece should be facing down. Measure 12½ inches (31.8 cm) up from the bottom, and attach another "L" shape in the same manner to create the finished hinges (photo a, previous page).

7. Repeat step 6 on the inside edge of the right frame, adding the remaining "L" shapes to the top and bottom of the frame.

Adding the Decorative Shapes

8. Cut and bend the rest of your found rod to create the inner rod shapes. There's nothing too exact about this step; have fun using your own creativity. When you've got the placement planned, weld the inner pieces to the frame pieces to create the screen design. Weld your geometric shapes in place (photo b).

9. Finish the project with spray finish or oil. Attach small felt pads to the bottom of the screen legs, if necessary, to protect the floors.

This mirror is just as much an art piece as it is useful. You create the contemporary brushed surface texture using a wire brush grinding disc.

Time to Reflect

Materials

80 inches (203.2 cm) of 2-inch (5 cm)
  flat bar
28 x 10-inch (71.1 x 25.4 cm) S
  scroll, ½ inch (1.3 cm) thick
Two 20 x 8-inch (50.8 x 20.3 cm) C
  scrolls, ½ inch (1.3 cm) thick
Eleven ½-inch (1.3 cm) ball bearings
12 inches (30.5 cm) of ¼-inch
  (6 mm) rod
14 x 24-inch (35.6 x 61 cm) mirror,
  ⅛ to ¼ inch (3 to 6 mm) thick

Tools & Supplies

MIG Welder Kit (page 11)
Angle Grinder Kit (page 12)
Clear acrylic spray
Glue

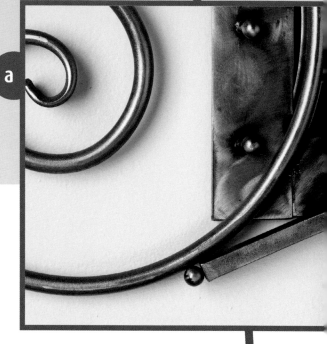

Time to Reflect

Step by Step

1. Measure, chalk, and cut the flat bar into two 13-inch (33 cm) and two 27-inch (68.6 cm) pieces.

2. To make the frame, lay out the cut flat bar pieces facedown on your worktable with the shorter pieces forming the top and bottom and the longer pieces forming the sides. The outside dimensions of the frame should measure 17 x 27 inches (43.2 x 68.6 cm).

3. Refer to the diagram on your machine for heat and wire speed settings, and then set your machine.

4. Weld the pieces together following the making a corner weld instructions on page 37. Clamp the frame down on the work surface, and smooth down the welds using the grinding disc.

5. Turn the frame over, and use the flap disc to grind a circular pattern onto the surface.

6. Using the photo as a guide, place the four curved pieces on the frame with the S scroll on the left side, the C scrolls at the top and bottom, and the double spiral on the right. Place short tack welds where the curved pieces touch the flat bar frame.

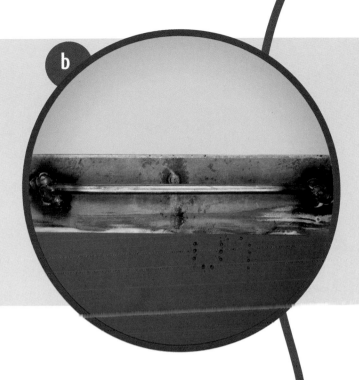

b

7. Turn the welded frame facedown, and make long tack welds on the back side to strengthen the connection between the curved pieces and the frame.

8. Turn the frame faceup, and weld ball bearings onto the ends of the C scrolls, using magnets or your gloved finger to brace the bearings. Continue adding bearings to the surface of the frame as desired, or as shown in photograph (photo a).

9. To make the hanging hook on the back, tightly secure the rod in your bench vise, and cut a 6-inch (15.2 cm) section. Bend each end of the cut piece slightly creating a very shallow "C" shape.

10. Place the frame facedown on your worktable, and lay the bent rod piece horizontally across the top, making sure that it stands approximately ¼ inch (6 mm) away from the frame surface. Hold the piece in place with magnets or a gloved finger, and place long tack welds on both ends of the bent rod. Make sure the welds are strong enough to hold up the weight of your mirror (photo b).

11. In order for the mirror to stand away from the wall an equal distance on the top and bottom, cut two more 3-inch (7.6 cm) pieces of rod, and bend them both to create a wide "V" shape. Weld the pieces onto the bottom left and right of the frame with the point of the "V" sticking out.

12. Turn the frame faceup, and go over the scroll pieces with a wire brush grinding disc to create a brushed surface.

13. Spray the frame with clear acrylic before gluing the mirror onto the back.

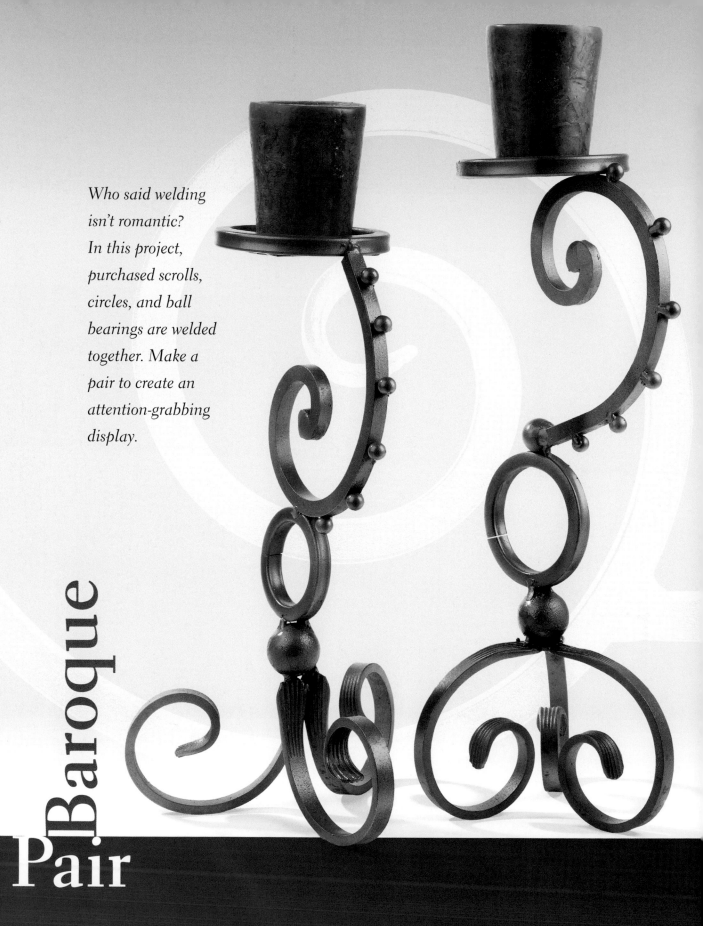

Who said welding isn't romantic? In this project, purchased scrolls, circles, and ball bearings are welded together. Make a pair to create an attention-grabbing display.

Baroque Pair

Materials (for large candleholder)

Three 5-inch (12.7 cm) scrolls
1-inch (2.5 cm) ball bearing
7-inch (17.8 cm) scroll
Six ⅜-inch (9.5 mm) ball bearings
Two 5-inch (12.7 cm) steel tubing
    rings from ½-inch (1.3 cm) square
    tubing
½-inch (1.3 cm) ball bearing
4 inches (10.2 cm) of 3-inch (7.6 cm)
    flat bar

Tools & Supplies

MIG Welder Kit (page 11)
Angle Grinder Kit (page 12)
Small magnet
Level, 12 inches (30.5 cm) or longer
Hammered texture spray paint

Step by Step

1. To get started, you'll need to first make a jig following the instructions in the sidebar on page 75. The jig will allow the legs to be placed and welded at the correct angle in order to build the rest of the candleholder on top.

2. Refer to the diagram on your machine for heat and wire speed settings, and then set your machine.

3. Clamp one of the 5-inch (12.7 cm) scrolls to one of the flat bar pieces on the jig. Clamp a second scroll to the jig in the same way, and place a long tack weld where the two scrolls touch. Repeat with the third scroll, and add more long tack welds where all three pieces touch.

4. Place the large ball bearing on top of where the three scrolls meet. Because the ball bearing is so thick, you will need to increase your heat and wire settings. Weld a large spot underneath where the ball touches the scrolls (photo a).

5. Chalk 1-inch (2.5 cm) increments along the length of the 7-inch (17.8 cm) scroll. Using a magnet, brace one of the small ball bearings on the first chalk mark, and weld it in place. Continue with this process until you've attached one ball bearing at each chalk mark.

9. Lay your 4-inch (10.2 cm) length of flat bar across the bottom of the 5-inch (12.7 cm) tubing ring, and weld where they touch. Hold the ring against the scroll so the middle of it is close to the center of the base, and weld it in place, using the level to make sure it's horizontal (photo b).

10. Spray the candleholder with the textured spray paint. Repeat steps 3 through 10 to create the shorter candleholder, turning the scrolls upside down and using a smaller middle tubing ring as shown in the picture.

6. Center one 5-inch (12.7 cm) tubing ring vertically on top of the ball bearing, and weld it in place.

7. Weld a medium-size ball bearing on top of the tubing ring.

8. Place the candle base on your worktable. Hold the 7-inch (17.8 cm) scroll up in the air so it's centered over one of the scrolls at the base. Before you weld this on, visualize where you'll attach the top 5-inch (12.7 cm) tubing ring—this ring will hold the candle on top—to make sure you have the scroll angled correctly. This step can be a bit tricky, so be patient and give yourself time to think it through. Once you've determined that you have the scroll angled correctly, weld it in place on top of the ball bearing.

Making a Welding Jig

Materials

16 x 16-inch (40.6 x 40.6 cm) piece of sheet or scrap metal

2 feet (61 cm) of 3-inch (7.6 cm) flat bar or scrap metal

14 inches (35.6 cm) of ⅜- to ¾-inch (9.5 mm to 1.9 cm) rod

1. Begin by laying out the piece of sheet metal, and draw a 12-inch (30.5 cm) circle on it with chalk. Place a chalk mark in the center of the circle and then, envisioning a clock face, draw another at the 12, 4, and 8 o'clock positions. Draw a line from the center of the circle to each of the perimeter marks—imagine cutting a pie into three pieces.

2. Place the flat bar in your bench vise. Measure and cut three 8-inch (20.3 cm) pieces.

3. Using a magnet, prop up a cut flat bar piece along one of the lines, making sure it stops in the center of the circle, and place four to eight short tack welds along both sides of the flat bar. Follow the same process with the other two flat bar pieces, welding them to the sheet metal.

4. Weld the 14-inch (35.6 cm) rod so it's sticking up from where the three pieces meet in the center of the circle.

When you're not cozying up to the fireplace, it's time to put this attractive screen on display. Adding fabric behind the spirals invites a myriad of design possibilities.

Cavorting Spirals
Screen

Materials

86 inches (218.4 cm) of ¾ x ¼-inch (1.9 cm x 6 mm) flat bar

36 inches (91.4 cm) of ⅜-inch (9.5 mm) rod

32 inches (81.3 cm) of ½-inch (1.3 cm) round tubing

Four 8-inch (20.3 cm) spirals

Seven 4-inch (10.2 cm) spirals

Three 5-inch (12.7 cm) spirals

Tools & Supplies

MIG Welder Kit (page 11)

Angle Grinder Kit (page 12)

Clear acrylic spray

Step by Step

1. Measure, chalk, and cut the flat bar into one 36-inch (91.4 cm) and two 25-inch (63.5 cm) lengths. Secure each piece in the bench vise, and smooth the cut edges with the grinding disc and your angle grinder.

2. Tightly secure the rod in the bench vise, and measure, chalk, and cut a 36-inch (91.4 cm) length just as you did with the flat bar. Tightly secure the tubing in the bench vise, and cut two 16-inch (40.6 cm) pieces.

3. Lay the flat bar pieces out on the welding table with the long bar at the top and the two 25-inch (63.5 cm) rods at the sides. Lay the rod across the bottom and make sure all the sides are touching. Brace the pieces in place with bricks.

4. Refer to the diagram on your machine for heat and wire speed settings, and then set your machine.

5. Put on your helmet and make short tack welds along each joint where the sections are touching. Go back over each short tack weld with two or three long tack welds (photo a).

6. With the frame lying right side up, position each spiral inside the rectangle frame, using the project photo as a guide. Place a large spot weld where each spiral touches the frame or another spiral (photo b). Turn the screen over and repeat this step on the backside.

TIP

If the spiral pieces you'll be welding inside the frame are thicker than the frame itself, you'll need to put something under the frame to hold it above the worktable about ¼ inch (6 mm) or so; I used grinding discs, placing one under each corner of the frame.

7. Use bricks to brace the screen up at a 90° angle to the worktable, and position one 16-inch (40.6 cm) length of tubing "foot" centered and perpendicular to the screen (photo c). Place a long tack weld where the foot touches the screen. Repeat this process for the other foot.

8. Coat the screen with clear or textured acrylic spray to guard against rusting.

Store your
vino in this
space-saving
wine rack. Its
compact design
—made from
flat bar and
purchased
grape leaves—
is a pleasant
combination
of form and
function.

Vine
Rack

Materials

12 feet (3.7 m) of 1¼ x ⅛-inch (3.2 cm x 3 mm) flat steel bar

Twenty-four 5-inch (12.7 cm) steel tubing rings from ½-inch (1.3 cm) square tubing

Twenty-four ⅜-inch (9.5 mm) ball bearings

Six 5 x 8-inch (12.7 x 20.3 cm) C scrolls

Six 3- to 4-inch (7.6 to 10.2 cm) cast or forged grape leaves

Tools & Supplies

MIG Welder Kit (page 11)
Angle Grinder Kit (page 12)
Drill and ¼-inch (6 mm) drill bit
Small magnet
Clear acrylic spray

Vine Rack

Step by Step

1. Measure, chalk, and cut your flat steel bar to create four 3-foot (0.9 m) lengths. Tightly secure each piece in the bench vise, and use your flap disc to create a brushed finish and to soften the cut edges.

2. Drill a ¼-inch (6 mm) hole in the center of each end of one piece, about ¼ inch (6 mm) in from the end. Repeat with the second piece, and then drill a hole in only one end for the last two pieces. These holes will allow you to attach the wine holder to the wall.

3. Refer to the diagram on your machine for heat and wire speed settings, and then set your machine.

4. Lay out one piece with two holes and one with one hole end to end on your worktable (using the photo as a guide), and weld the pieces together. Repeat with the other two flat bar pieces.

5. Lay the two welded pieces side by side on your worktable, and make the following small chalk marks starting from one end: at 1 inch (2.5 cm), 6 inches (15.2 cm), 7 inches (17.8 cm), 12 inches (30.5 cm), and so on until you have twelve

5-inch (12.7 cm) spaces with 1 inch (2.5 cm) between them. Next, make a large chalk mark in the middle of each 5-inch (12.7 cm) space; this is where you will be welding on the tubing rings.

6. With the two pieces of flat bar spaced evenly apart on your worktable, lay the scrolls flat between the bars, taking care to evenly space them as shown in the photograph. Weld them in place where they touch with long tack welds.

7. Using the small magnet and starting from one end, prop up a 5-inch (12.7 cm) ring on the flat bar at a large chalk mark. Make a short tack weld on both sides of the ring to set it in place, and go over the first welds with long tack welds for reinforcement, making sure you don't melt a hole in the ring (photo a). Continue adding all the rings on both flat bar pieces—you should end up with 12 on each bar—making sure they are directly across from each other.

8. Attach the grape leaves to the flat bar and the scrolls, using the photo as a guide.

9. Using a small magnet, brace a ball bearing on top of each ring, and place a long tack weld on one side where it touches (photo b).

10. Coat the entire rack with clear acrylic spray to guard against rusting.

Need a new side table for the porch? This go-anywhere piece— made completely of architectural steel parts—makes a fine platform for a cup of tea or a blooming plant.

Table for Two

Materials

Three 20-inch (50.8 cm) C scrolls
10-inch (25.4 cm) steel tubing ring
  from ½-inch (1.3 cm) square
  tubing
5-inch (12.7 cm) steel tubing ring
  from ½-inch (1.3 cm) square
  tubing
Twenty-one ⅜-inch (9.5 mm) ball
  bearings
18-inch (45.7 cm) round glass
  tabletop, ¼ inch (6 mm) thick

Tools & Supplies

MIG Welder Kit (page 11)
Angle Grinder Kit (page 12)
Level, 12 inches (30.5 cm) or longer
Clear acrylic spray
Small felt pads (optional)

Step by Step

1. First make the jig, following the sidebar on page 75.

2. To start building the table, clamp a scroll to one of the flat bar pieces on the jig.

3. Clamp a second scroll in the same way, and lay your level across the top of both pieces, making adjustments until they're level across the top. Refer to the diagram on your machine for heat and wire speed settings, and then set your machine. Place a long tack weld where the spirals touch.

4. Clamp the third scroll onto the jig—checking everything once again with your level—and add more long tack welds where all three pieces touch.

5. Place the 10-inch (25.4 cm) ring on top, and, after using your level to make sure it is straight, tack weld the ring in place where it touches the scrolls.

6. Turn the table upside down, and place the 5-inch (12.7 cm) ring inside the scroll legs, making sure the ring is level. Place tack welds where it touches the scrolls (photo a).

7. Working down from the 5-inch (12.7 cm) ring, make six chalk marks—placed 2 inches (5 cm) apart—along the inside curve of each scroll.

8. Using a 1-inch (2.5 cm) magnet as a brace, hold a ball bearing against the scroll at one of the chalk marks, and weld it in place. Continue until you have finished welding six ball bearings onto each scroll.

9. Add a ball bearing to the end of each scroll leg, again using the 1-inch (2.5 cm) magnet for bracing.

10. Spray the table with clear acrylic spray to guard against rusting.

11. Turn the table over and lay the glass on top, inserting small felt pads where the glass and metal touch.

Showcase your
prized begonia
in this botanical-
inspired plant stand.
Found steel objects
give it an organic
freeform style, just
right for a relaxing
porch setting.

Botanical
Stand

Materials

3 found narrow legs, between 20
   and 40 inches (50.8 and 101.6 cm)
   long
10-inch (25.4 cm) ring
3 found side pieces, each longer
   than 16 inches (40.6 cm)
Six ½-inch (1.3 cm) ball bearings
   (for decoration)
Several found metal objects
   (for decoration)

Tools & Supplies

MIG Welder Kit (page 11)
Angle Grinder Kit (page 12)
Clear acrylic spray or
   oil

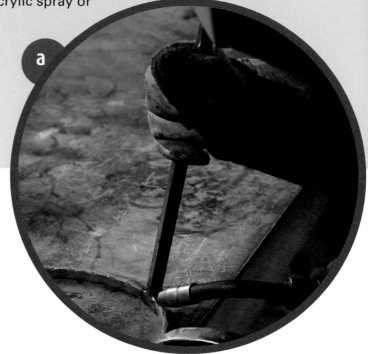

Botanical Stand

Step by Step

Preparing the Pieces

1. If your found items are particularly rusty, clean each piece with the flap disc on your angle grinder. Measure and cut the legs; the shape of the legs can vary greatly as long as they all measure the same length.

2. Place the ring on your worktable, draw a triangle in the middle of the circle, and make a chalk mark on the ring at each point on the triangle. These marks will indicate where to attach the legs.

3. Refer to the diagram on your machine for heat and wire speed settings, and then set your machine.

Attaching the Legs

4. Working with the piece upside down on your table, hold one leg up at the desired angle so it rests against the ring. Place a short tack weld where the leg touches the ring (photo a).

5. Follow the same steps to attach your second leg at the second chalk mark on your ring. Angle the leg so that, when the piece is flipped right side up, the legs will be about 12 inches (30.5 cm) apart at the bottom.

6. Turn the piece right side up. Before welding on the third leg, make sure the two legs are at the correct angle so the ring is horizontal. With the piece still right side up, slide the third leg under the ring and place a short tack weld where they touch. When you are satisfied that the legs are correct, turn the stand over and place two or three long tack welds at each joint for reinforcement.

b

TIP

If you need to make adjustments, you can easily break the weld by hitting it hard on the worktable or by grinding off a small amount of the tack weld and breaking it apart.

Adding the Side Pieces

7. Hold up the side pieces, and place chalk marks where you'll be welding them onto the legs. If your legs are rusty where you want to attach the side pieces, grind off a 1-inch (2.5 cm) spot before attaching them.

8. Place short tack welds to connect the side pieces to the legs (photo b). When you're certain the side pieces are in the right place, go over each short tack weld with two or three long tack welds to strengthen them.

9. Add ball bearings and other decorations, and then grind your piece until you get the desired finish. To protect your piece from rust, coat each surface with acrylic spray or oil.

Inspired by nature,
new life is given
to scrap metal in
the form of a coat
stand. Branch and
twig-like pieces
provide handy
hanging spots
to accommodate
everything from
jackets to purses.

Family
Tree

Materials

Three 5- to 6-foot-long (152.4 to 182.9 cm) pieces of ⅜-inch (9.5 mm) or thicker scrap rod
5 feet (152.4 cm) of ¼-inch (6 mm) scrap rod

Tools & Supplies

MIG Welder Kit (page 11)
Angle Grinder Kit (page 12)
Clear acrylic spray or oil

Step by Step

1. Measure, chalk, and cut the thicker rod so you have three 5- to 6-foot-long (152.4 to 182.9 cm) pieces. Cut the thinner rod into 6- to 13-inch (15.2 to 33 cm) sections to create the number of hooks you'd like the coatrack to have.

2. Bend the rod with your bending tube, using the photo as a guide.

3. Refer to the diagram on your machine for heat and wire speed settings, and then set your machine.

4. Working with two pieces of rod, place your grounding clamp onto one of the pieces, hold them both up, and weld them together where you think they should be joined (photo a). If where you'll be welding is rusty, clean it by grinding a 1-inch (2.5 cm) spot with the flap disc. Make sure the legs—where the coatrack will be touching the floor—are even and at least 12 inches (30.5 cm) apart.

b

5. Hold up the third piece and weld it where it touches the other two, making certain that the base forms a triangle—or tripod—with each leg the same distance apart.

TIP

If your rod is too thin to create a stable base, weld side pieces across the bottom of your coatrack. See the Botanical Stand (page 85) or Sitting Pretty (page 118).

6. Bend the thinner rod pieces to make hooks, and weld them on wherever you'd like them to be (photo b).

7. Finish the whole coatrack with the wire wheel brush on your angle grinder until you get the look you want, and then use acrylic spray or oil finish to protect the piece from rust.

A bold geometric
design shows this
wall sconce off in
a favorable light.
When illuminated,
the pierced shapes in
the metal project a
contemporary design.

Geometric
Sconce

Materials

6 x 24-inch (15.2 x 61 cm) piece of 18-gauge flat sheet

12 x 18-inch (30.5 x 45.7 cm) piece of 18-gauge flat sheet

10 inches (25.4 cm) of 2 x ⅛-inch (5 cm x 3 mm) flat bar [A]

18 inches (45.7 cm) of ⅛-inch (3 mm) rod

15 inches (38.1 cm) of 1 x ¾-inch (2.5 x 1.9 cm) flat bar [B]

20 inches (50.8 cm) of 1 x ⅛-inch (2.5 cm x 3 mm) flat bar [C]

Tools & Supplies

MIG Welder Kit (page 11)

Angle Grinder Kit (page 12)

Ball hammer

Screwdriver

Drill and ¼- and ½-inch (0.6 and 1.3 cm) drill bits

2 x 4-inch (5 x 10.2 cm) piece of wood

Basic lamp-making kit

Geometric Sconce

Step by Step

1. Measure, chalk, and cut the 6 x 24-inch (15.2 x 61 cm) piece of flat sheet metal in half so you end up with two 6 x 12-inch (15.2 x 30.5 cm) pieces. Use the grinding disc to smooth the cut edges of the metal.

2. Secure one 6 x 12-inch (15.2 x 30.5 cm) piece in the bench vise, and score three 2-inch (5 cm) slits in the center for light to shine through. Use the same steps to cut three slits in the second 6 x 12-inch (15.2 x 30.5 cm) piece and 14 slits in the 12 x 18-inch (30.5 x 45.7 cm) piece of flat sheet (photo a).

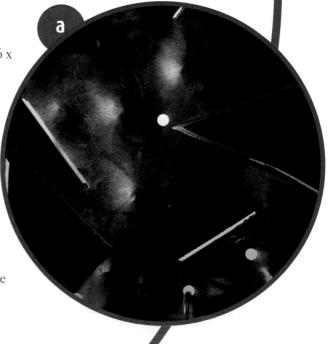

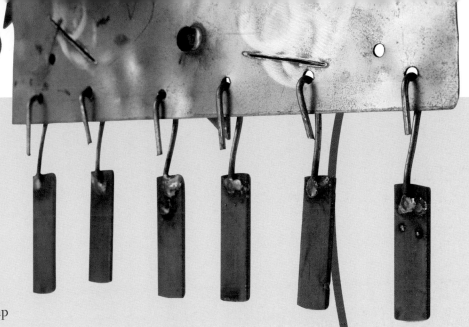

3. Go over all three pieces with the flap disc using a small circular motion to create a brushed and finished surface.

Creating the Triangles

4. Secure flat bar A in the bench vise, and then measure, chalk, and cut a 2-inch (5 cm) square from the end. Chalk and cut a diagonal line across the square so you end up with two triangle pieces. Repeat this process until you have nine triangle pieces.

5. Lay the 12 x 18 inch (30.5 x 45.7 cm) piece of metal on the wood, finished side down, and pound it on the back with the ball hammer to create a uniform textured finish. The steel will curl up during this process; to straighten it, lay the piece faceup on your work surface with a leather glove on top, and pound it flat with the hammer. The glove will help to protect the metal while you work to straighten it out.

6. Drill light holes in the surface of the 12 x 18-inch (30.5 x 45.7 cm) piece using a ¼-inch (6 mm) drill bit. Drill nine place holes using the ½-inch (1.3 cm) drill bit; these will serve as welding places for the triangle pieces, since you'll be welding from behind the piece.

7. Position your triangle pieces on the worktable under the lamp face, lining up each ½-inch (1.3 cm) hole with each triangle piece. When you've got the positioning just right, place a weld inside each hole. Continue this process until all the triangle pieces are welded on from behind (photo c).

Making the Fringe

8. Create a bend in the ⅛-inch (3 mm) rod 1-inch (2.5 cm) from end. Cut the rod 2 inches (5 cm) from the bend, and continue this process until you have six hooks.

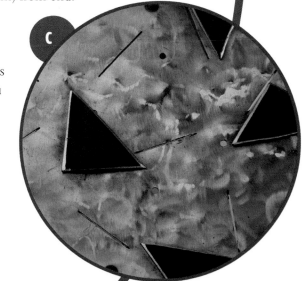

d

Finishing the Back

11. Lay the lamp facedown on your worktable. Hold one 6 x 12-inch (15.2 x 30.5 cm) flat sheet piece perpendicular to the lamp face, making sure it is 1 inch (2.5 cm) in from the side edge and 3 inches (7.6 cm) in from the top and bottom. Place five or six short tack welds on the inside joint. Repeat this process with the other 6 x 12-inch (15.2 x 30.5 cm) piece on the opposite side.

12. Measure, chalk, and cut two 10-inch (25.4 cm) sections from flat bar C. Drill a ½-inch (1.3 cm) hole in the center of both pieces. Weld one piece across the bottom of the side pieces, and weld the other piece—which will serve as a nail hole—across the top (photo d).

13. Install the parts from the lamp kit.

9. Working along the bottom edge of the lamp face, measure and chalk six holes, each about 2 inches (5 cm) apart and ¼ inch (6 mm) from the bottom. Drill the holes using the ¼-inch (6 mm) drill bit.

10. Go over the surface of flat bar B with your flap disc. Measure, chalk, and cut six 2½-inch-long (6.4 cm) slices, and then grind the cut edges smooth. Hold each piece down by placing a brick on one end, and weld the hooks onto the back of each piece. Hang the hooks through the holes (photo b).

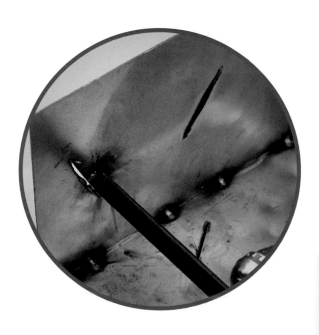

With a little creativity and some ornate found steel pieces, you can create this charming accent table that seems to be planted where it stands.

Harvest Table

Materials

11-inch (27.9 cm) half-circle piece of
flat steel, no more than ¼-inch
(6 mm) thick (for the tabletop)
Three leg pieces, each 27 inches
(68.6 cm) long
Three side pieces, each 10 to 14
inches (25.4 to 35.6 cm) long

Tools & Supplies

MIG Welder kit (page 11)
Angle Grinder kit (page 12)
Clear acrylic spray

Harvest Table

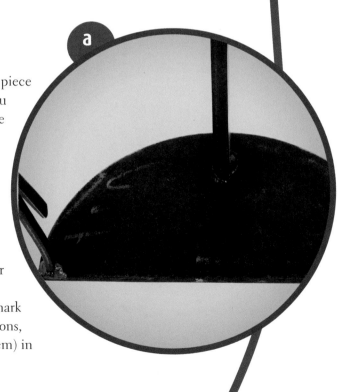

Step by Step

1. Clean and finish each found piece
with your angle grinder until you
get the look you want and you've
got enough clean spots to place
your welds. Chalk and cut all
your pieces to the dimensions
noted in the materials list.

2. Lay your tabletop piece
facedown on the worktable.
Make a chalk mark in the center
of the half circle, and then,
picturing a clock face, make a mark
at the 12, 3, and 9 o'clock positions,
working 1 to 2 inches (2.5 to 5 cm) in
from the outer edge.

3. Refer to the diagram on your machine
for heat and wire speed settings, and then
set your machine.

4. Hold one leg against the table at one of the chalk marks. To make sure you get the correct angle, envision drawing an imaginary triangle in the air with the ends of your legs, with the legs an equal distance apart from one another. When you're satisfied with the angle of the leg, short tack weld it to the tabletop. Repeat this step for the other two legs (photo a).

5. Turn the table over and check the angles of the legs. Once you're satisfied, turn the table upside down again, and place two to four long tack welds at each joint for reinforcement.

6. Turn the table right side up again, and weld the side pieces to the table legs. For a sturdy table, weld the side pieces within the bottom 6 inches (15.2 cm) of the legs.

7. Finish the table as desired with your soft flap disc or your wire brush. Coat the table with clear acrylic spray to guard against rusting.

Make a statement in your dining room or entryway with a dramatic vase holder. Created with purchased architectural parts, this piece is as easy to make as it is to admire.

Tripod Vase

Materials

Three 28 x 10-inch (71.1 x 25.4 cm) S scrolls, ¼ inch (6 mm) thick

Two 5-inch (12.7 cm) steel tubing rings from ½-inch (1.3 cm) square tubing

4¼-inch (10.8 cm) circular candle drip pan

Twenty-four ⅜-inch (9.5 mm) ball bearings

4 x 13-inch (10.2 x 33 cm) cylinder glass vase

Tools & Supplies

MIG Welder Kit (page 11)
Angle Grinder Kit (page 12)
Small magnets
Level (optional)
Clear acrylic spray or oil

Step by Step

1. Make the jig, following the sidebar on page 75.

Joining the Legs

2. Clamp two of the S scroll pieces to the jig, leaving enough room at the top for one of the 5-inch (12.7 cm) tubing rings. Refer to the diagram on your machine for heat and wire speed settings, and then set your machine. Place a short tack weld where the scrolls touch.

3. Unclamp the pieces from the jig, and place them on the worktable. Hold the third S scroll against the joined pieces with one hand, making sure the tubing ring will fit between them. Weld the third S scroll where it touches the first two (photo a).

Adding the Rings

4. Hold the top tubing ring horizontally—you might want to use a level—between the three scrolls, and weld it in place where it touches the scrolls.

5. Turn the scrolls upside down, and weld the bottom tubing ring in place between the scrolls. To check for alignment, try holding a rod vertically through the two circles; these two rings must be closely aligned to hold the glass vase.

6. While the vase holder is upside down, center the drip pan over the bottom circle, and weld it where it touches the edges (photo b).

7. Turn the vase holder right side up and, using magnets to brace the ball bearings, weld the bearings in place along each scroll (photo c) and where the scrolls are welded together, using the photo as a guide.

8. Coat the vase with clear acrylic spray to prevent rust. Insert the glass vase.

Advanced Projects

Display this graceful console table by itself or with the matching mirror shown on page 69. It's made with all architectural pieces so no cutting or bending is required.

Better
Half

Materials

Thirteen ½-inch (1.3 cm) ball
  bearings
Four 20-inch-long (50.8 cm) square
  C scrolls, ½ inch (1.3 cm) thick
28-inch-long (71.1 cm) flat S scroll,
  ½ to ¾ inch (1.3 to 1.9 cm) thick
Three 28-inch-long (71.1 cm) round
  S scrolls, ½ to ¾ inch (1.3 to 1.9
  cm) thick
30-inch (76.2 cm) half-circle glass
  tabletop, ½ inch (1.3 cm) thick

Tools & Supplies

MIG Welder Kit (page 11)
Angle ruler
Clear rubber pads
Clear acrylic spray

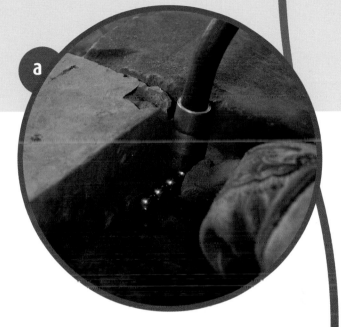

Step by Step

Creating the Tabletop

1. Refer to the diagram on your machine
for heat and wire speed settings, and then
set your machine. Start by welding a ball
bearing to the end of one C scroll, using a
small magnet or brick for bracing.

2. Lay out the flat S scroll on your
worktable and butt it up against the C
scroll piece, with both pieces facedown.
Weld the pieces together where they
touch.

Constructing the Legs

3. With the larger spiral at the bottom,
lay out two of the big S scrolls so they're
facing each other. Measuring from
outside to outside, the smaller spirals,
at the top, should be approximately 24
inches (61 cm) apart, and the larger
spirals, at the bottom, should be about 26
inches (66 cm) apart.

4. Lay a C scroll between the two scrolls,
and place a short tack weld where it
touches the big S scrolls. When you are
sure of the placement, go back and secure
the welds with long tack welds. This piece
will form the back of the table.

5. Prop up two ball bearings
alongside a brick, and weld
them together where they
touch. Add another ball
bearing, and weld it to the
end of the first two. Repeat
for the fourth ball bearing.
Do the same thing until you
have three "ball bearing rods,"
each made up of four bearings
(photo a).

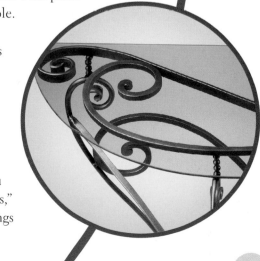

6. Weld the ball bearing rods on top of the two attached S scrolls, making sure they are completely vertical (photo b). Do the same with the remaining S scroll.

Attaching the Legs

7. Butt the tabletop pieces against the ball bearing rods, using bricks or a gloved hand for bracing. Make sure you are holding the tabletop at a 90° angle to the legs by checking the angle ruler.

8. Place short tack welds where the tabletop touches the ball bearings. Check the angles, and then go back over the short tack welds with long tack welds.

9. Stand your table up and balance the other S scroll under the table, so the ball bearing rod touches the C scroll on the tabletop. Using the level, get the tabletop as horizontal as possible. Place a short tack weld where the third leg touches the tabletop. Go over the first weld with a long tack weld when you are certain it is exactly where you want it.

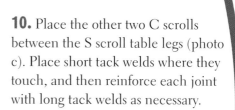

10. Place the other two C scrolls between the S scroll table legs (photo c). Place short tack welds where they touch, and then reinforce each joint with long tack welds as necessary.

11. Coat the entire table with clear acrylic spray. Place the glass tabletop on top of the frame, using clear rubber pads for padding where the glass and metal touch.

Small leaves add a natural yet chic feel to this understated chandelier. Glass votive holders rest inside metal circles, creating a floating candleholder.

Flora
Lumina

Materials

52½ inches (133.4 cm) of ⅜-inch (9.5 mm) rod

5 inches (12.7 cm) of ³⁄₁₆-inch (5 mm) rod

5 steel rings, with 2-inch (5 cm) inside diameters

Thirty to forty 1-inch (2.5 cm) steel leaves

4 feet (121.9 cm) of chain

5 fluted glass votive holders, 2 inches (5 cm) in diameter

Tools & Supplies

MIG Welder Kit (page 11)
Angle Grinder Kit (page 12)
Small level (optional)
Pliers
Clear acrylic spray

Step by Step

1. Measure, chalk, and cut five 10½-inch (26.7 cm) pieces of the thicker rod and two 2½-inch (6.4 cm) pieces of the thinner rod.

Bending the Armatures

2. Working on one 10½-inch (26.7 cm) piece at a time, place the end in the bench vise, and, using a bending tube, make a small bend. Using ¼-inch (6 mm) increments, continue this process until you've created an "L" shape. Repeat this process for the other four rod pieces, using the first one as the template.

3. To create a template for the armature placement, draw a 10-inch (25.4 cm) circle on your welding table. Inside the circle, draw five lines to create a star shape and mark a dot in the center of the circle. Draw five straight lines radiating out from the dot in the center through each point of the star.

4. Refer to the diagram on your machine for heat and wire speed settings, and then set your machine.

Joining the Armatures

5. Using bricks, prop up two bent rod pieces so they are standing vertically, with each piece radiating out from the center along two of the lines of the star. Hold them with one hand, and short tack weld where they touch (photo a).

6. Brace a third bent rod along another radiating line, and short tack weld it where it touches the first two pieces. Repeat this process until you have all five rods welded together, and then add another four or five long tack welds at each joint for reinforcement.

7. To create the hanging rod, secure one of the 2½-inch (6.4 cm) rod pieces in the bench vise, and use the bending tube to bend it into a "V" shape, until the ends are about 1 inch (2.5 cm) apart. Hold the hanging rod piece with the point of the "V" facing up, and weld the two ends to the top of your chandelier base (photo b).

Adding the Rings

8. To weld on the rings that will hold the glass votive holders, hold one ring against the end of an "L," eyeing the ring or using a level to make certain it's horizontal. Place a short tack weld where they touch. Check the angle once again, and place a long tack weld to reinforce it. Attach the other four rings in the same manner.

9. Weld on your leaves as desired.

10. Secure the end of the second 2½-inch (6.4 cm) rod piece in the bench vise, and, using a hammer, bend it in half to create the chain link that will fasten the chain to the chandelier (photo b). Pull the chain link through the hanging piece and the end of the chain to connect the two. Use pliers to close the link. Tack weld the link together for reinforcement.

11. Coat the chandelier with clear acrylic spray to prevent rust.

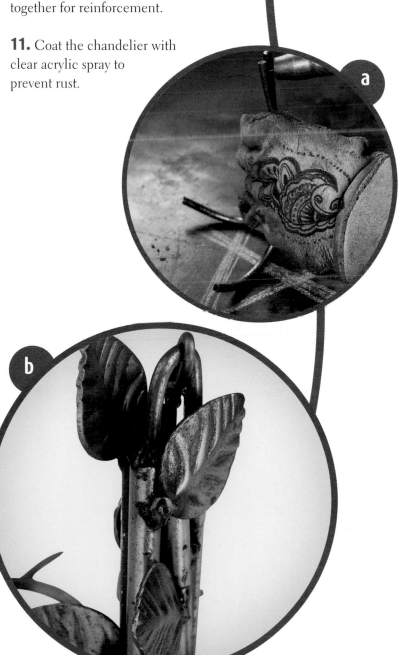

Do you treat your four-legged friend like a king or queen?
Canines and felines alike will feel like royalty atop this piece
made from purchased architectural steel and a padded seat.

Pet Throne

Materials

35 inches (88.9 cm) of 1 x ⅛-inch (2.5 cm x 3 mm) flat bar

Four 5-inch (12.7 cm) finials (for the legs)

4 feet (121.9 cm) of ½-inch (1.3 cm) round rod

Twenty ½-inch (1.3 cm) ball bearings

Two 1-inch (2.5 cm) ball bearings

2 scrolled fish tail ends, about 11¾ inches (29.8 cm) long

2-inch (5 cm) ball bearing

Round wooden tabletop, 18 inches (45.7 cm) in diameter

Foam, 18 inches (45.7 cm) in diameter

1 yard (0.9 m) of fabric

Tools & Supplies

MIG Welder Kit (page 11)

Angle Grinder Kit (page 12)

Drill and ¼-inch (6 mm) drill bit

Small magnets

Square ruler (optional)

Clear acrylic spray or oil

Staple gun

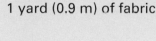

a

Step by Step

1. Measure, chalk, and cut two 13-inch (33 cm) pieces and one 9-inch (22.9 cm) piece from the flat bar.

Building the Base

2. Lay the pieces on your worktable to form an "H" shape, and weld where the pieces touch each other (photo a). Working on the 13-inch (33 cm) pieces, drill a ¼-inch (6 mm) hole 1 inch (2.5 cm) in from each end until you have four holes; this is where the wooden seat will be drilled in.

3. Making sure you've got them angled correctly—using a square ruler, if necessary—weld the finial feet onto the ends of the 13-inch (33 cm) pieces, between the hole you just drilled and the end (photo b).

Creating the Crown
4. Measure, chalk, and cut four 12-inch (30.5 cm) pieces of rod. Lay each piece of rod on your worktable, brace the ½-inch (1.3 cm) ball bearings in place with small magnets, and weld them onto the rod, using the photo as a guide.

5. Weld the 1-inch (2.5 cm) ball bearings onto the fish tail end pieces (photo c). Because the scroll on the fish tail end doesn't allow it to lie flat, you might need to prop it up so that it's centered in the middle of the ball bearing.

6. Assemble the fish tail end pieces and rod pieces on your worktable to resemble the crown shape (photo d). Brace and then weld the 2-inch (5 cm) ball bearing in place on the top center of the crown. You'll be welding the crown to the H-shaped base, so make sure the bottom of the fish tail end pieces line up with the ends of the base. Short tack weld the pieces of the crown together where they touch. Place long tack welds over the initial short tack welds to strengthen the piece.

7. Hold the crown perpendicular to the base, and weld it to the H-shaped base where they touch.

8. Coat the entire piece with clear acrylic spray or oil.

Making the Bed

9. Place the foam on the wood and then stretch the fabric over the foam. Smooth the fabric and then staple the ends to the backside of the wood. Turn the seat over and drill it in place through the holes you created in step 2.

The entryway, the bedroom, the screened-in porch— can't decide where to place this artful bench? The plush seat is easily covered and recovered to match any whim or any setting.

Sit-a-Spell
Bench

Materials

5 feet (152.4 cm) of 1 x ⅛-inch
(2.5 cm x 3 mm) flat bar
2 large S scrolls, about 28 inches
(71.1 cm) long
Large C scroll, about 20 inches
(50.8 cm) long
Four 1-inch (2.5 cm) ball bearings
Twenty ½-inch (1.3 cm) ball bearings
Four 2-inch (5 cm) ball bearings
4 scrolled fish tail ends, about 11¾
inches (29.8 cm) long
2 S scrolls, about 12 inches (30.5
cm) long
71 inches (180.3 cm) of plain round
½-inch (1.3 cm) bar
Wood for the seat, 14 x 34 inches
(35.6 x 86.4 cm)
4-inch-thick (10.2 cm) foam padding,
approximately 14½ x 34½ inches
(36.8 x 87.6 cm)
1½ yards (1.4 m) of fabric

Tools & Supplies

MIG Welder Kit (page 11)
Angle Grinder Kit (page 12)
Drill and ¼-inch (6 mm) drill bit
Small magnets
Angle ruler (optional)
Hammered texture spray paint
Scissors
Staple gun

Step by Step

1. Measure, chalk, and cut two 12-inch
(30.5 cm) pieces and one 28-inch (71.1
cm) piece of flat bar.

Creating the Frame

2. Make the base by laying the three cut
pieces out on your worktable in an "H"
shape, and weld them together where
they touch. Working on the 12-inch (30.5
cm) pieces, drill a ¼-inch (6 mm) hole 2
inches (5 cm) in from each end until you
have four holes; this is where the wooden
seat will be drilled in.

3. Create the back of the bench by arranging the two large S scrolls and the large C scroll on your worktable, using the photo as a guide. Make sure the bottoms of both scrolls are lined up with the ends of the "H"-shaped frame on the seat; this is where the back will be welded onto the base. Refer to the diagram on your machine for heat and wire speed settings, and then set your machine. Weld the pieces together where they touch (photo a).

4. Weld ball bearings along the S scroll pieces, using bricks to prop the large ball bearings and magnets to brace the small ones; evenly space and then attach four 1-inch (2.5 cm) ball bearings along the left S scroll and eight ½-inch (1.3 cm) ball bearings along the right.

Making the Legs

5. Butt a 2-inch (5 cm) ball bearing against one of your fish tail ends. Prop up the fish tail end until it's even with the middle of the ball, brace the pieces with a brick, and weld them together where they touch. Repeat this step until you have four legs.

Assembling the Parts

6. Lay the "H"-shaped frame upside down on your worktable. Hold one leg upside down against the end of the "H"; to make sure the leg is straight, you might want to use an angle ruler. Holding the end of the leg with one hand, place a short tack weld where it touches the frame. Look to see if the leg is straight from all angles, and, if not, break it off and weld it again. When you've got the leg straight, reinforce the short tack weld with long tack welds (photo b). Repeat this step until you have three legs welded on.

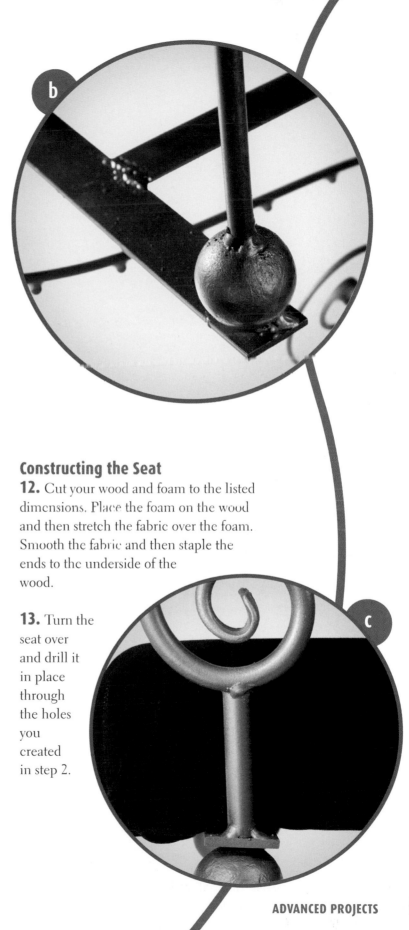

7. Turn the bench right side up, and position the fourth leg in place. Place a short tack weld where it touches the "H." Turn the bench upside down again, and finish welding the fourth leg.

8. Turn the bench right side up, and weld the S scroll pieces along the sides as shown.

9. Measure and cut the ½-inch (1.3 cm) bar to create one 35-inch (88.9 cm) piece and one 36-inch (91.4 cm) piece. Using the photo as a guide, bend the pieces slightly to make the cross pieces at the front and back of the bench, using the longer piece—which will create a more dramatic curve—for the back of the bench. Weld the front and back rod pieces to the legs where they touch.

10. Pick up the back piece and hold it so it's almost perpendicular to the ends of the bench base. If the back is welded at exactly a 90° angle, the bench will be uncomfortable to sit in. Weld the back to the base (photo c).

11. Spray the piece with the textured spray paint.

Constructing the Seat

12. Cut your wood and foam to the listed dimensions. Place the foam on the wood and then stretch the fabric over the foam. Smooth the fabric and then staple the ends to the underside of the wood.

13. Turn the seat over and drill it in place through the holes you created in step 2.

From the fields to your home, an old tractor seat finds new life as a chic chair. Asymmetrical legs are the perfect fit for this assemblage of found objects.

Sitting Pretty

Materials

Two metal leg pieces, about 18 inches (45.7 cm) long
Found tractor or chair seat
One 36-inch (91.4 cm) metal piece (to serve as chair back and leg)
Three narrow side pieces, about 14 inches (35.6 cm) long
Seven ⅜-inch (9.5 mm) ball bearings
Found metal shapes (for decoration)

Tools & Supplies

MIG Welder Kit (page 11)
Angle Grinder Kit (page 12)
Angle ruler
Clear acrylic spray or oil

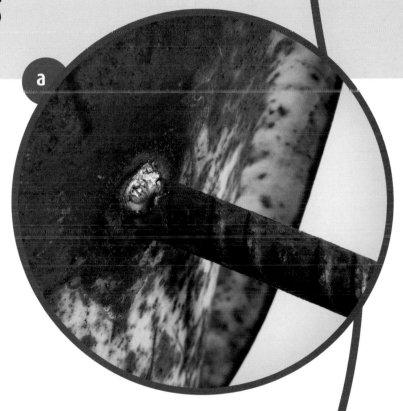

Step by Step

Preparing the Parts

1. Clean and finish each piece with your angle grinder until you have the look you want and you've got enough clean spots to place your welds.

2. Chalk and cut the two metal leg pieces to the desired height. Eighteen inches (45.7 cm) is standard for chair legs.

3. Refer to the diagram on your machine for heat and wire speed settings, and then set your machine.

Attaching the Legs

4. Turn the seat upside down, and attach the two legs with short tack welds. Step back and check the angle; depending on the shape of your pieces, you'll have to use an angle ruler or eye it to get the legs on straight. If it looks right, go over each short tack weld with a long tack weld for reinforcement (photo a).

5. Hold the chair upright and butt the back piece up against the back of the chair seat. When you have it level, place a tack weld where the back piece and the seat touch (photo b). Stand back and see if it is where you want it. When you are sure, place two or three long tack welds for reinforcement.

TIP

It's much easier to create chairs and tables with three legs than with four. Three legs are simple to stabilize while the addition of the fourth leg requires more precision!

Adding the Side Pieces

6. With the chair turned upright, weld on the three side pieces to get the look you want. Weld them onto the lower half of the legs to keep the piece sturdy.

7. Attach seven ball bearings along the front side piece, using small magnets to brace them in place while you weld them on (photo c).

8. Weld a small, disk-shaped found object to the bottom of the left leg (photo d). Weld a metal spiral or another metal shape onto the chair back.

9. Finish your chair as desired. To prevent rust, coat the piece with clear acrylic spray or oil.

b

Swirly Shelves

Hand-bent scrolls of varying lengths supply the support structure for these shelves.
The result is a wonderful, asymmetrical look.

Materials

12 feet (3.7 m) of ¼-inch (6 mm) rod
Two 6 x 12-inch (15.2 x 30.5 cm)
   pieces of flat steel
Seven ⅜-inch (9.5 mm) ball bearings

Tools & Supplies

MIG Welder Kit (page 11)
Angle Grinder Kit (page 12)
Clear acrylic spray

Step by Step

1. Measure, chalk, and cut lengths of rod to the following dimensions: one at 18 inches (45.7 cm), 16 inches (40.6 cm), 12 inches (30.5 cm), and three at 30 inches (76.2 cm).

Creating the Scrolls

2. Make an S scroll by placing a 30-inch (76.2 cm) piece in the bench vise ½ inch (1.3 cm) from the end. Tighten the vise. Bend the rod a little, and then continue feeding it through and bending it in ¼-inch (6 mm) increments until you're satisfied with the curve on one end. Turn the rod over and begin to bend it in the opposite direction, using the photo as a guide. Repeat this process with the other 30-inch (76.2 cm) cut rod.

3. Bend the 12-, 18-, and 30-inch (30.5, 45.7, and 76.2 cm) pieces of rod so each has a spiral on just one end (photo a).

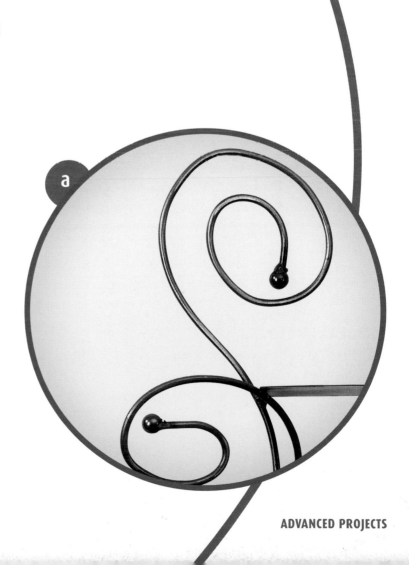

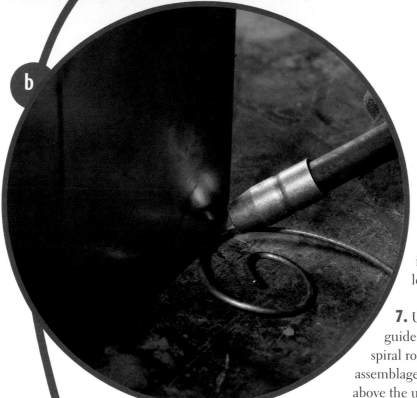

6. Place the right back corner of the upper shelf on top of the right S scroll. Make sure the shelf is level, and weld it in place using short and then long tack welds.

7. Using the project photo as a guide for placement, butt a single spiral rod against the left side of the assemblage so that the spiral juts up above the upper shelf and the lower end of the rod touches the S scroll a few inches above the lower shelf. Position the rod against the other pieces so you've got a small hole for a screw (photo c). Make a short tack weld where the rod pieces meet and where the rod meets the upper shelf. After making sure the upper shelf is level, go back over the short tacks with long ones.

Adding the Shelves

4. Lay an S scroll on the worktable. Measure at about the center point, and hold one of the flat steel pieces against it at a 90° angle. This will become the lower shelf. Refer to the diagram on your machine for heat and wire speed settings, and then set your machine. Place short tack welds where the two pieces touch (photo b). When you're certain the placement is correct, go over the short tack welds with long tack welds.

5. Lay the other S scroll on your worktable so it's a mirror image of the first one, then move the whole S scroll up about 3 to 4 inches (7.6 to 10.2 cm) higher than the one on the left. This will give your piece a whimsical, asymmetrical look. Butt the shelf piece up to the right S scroll, make sure the shelf is level, and place short tack welds where the pieces touch. When you're certain the placement is correct, go over the short tack welds with long tack welds.

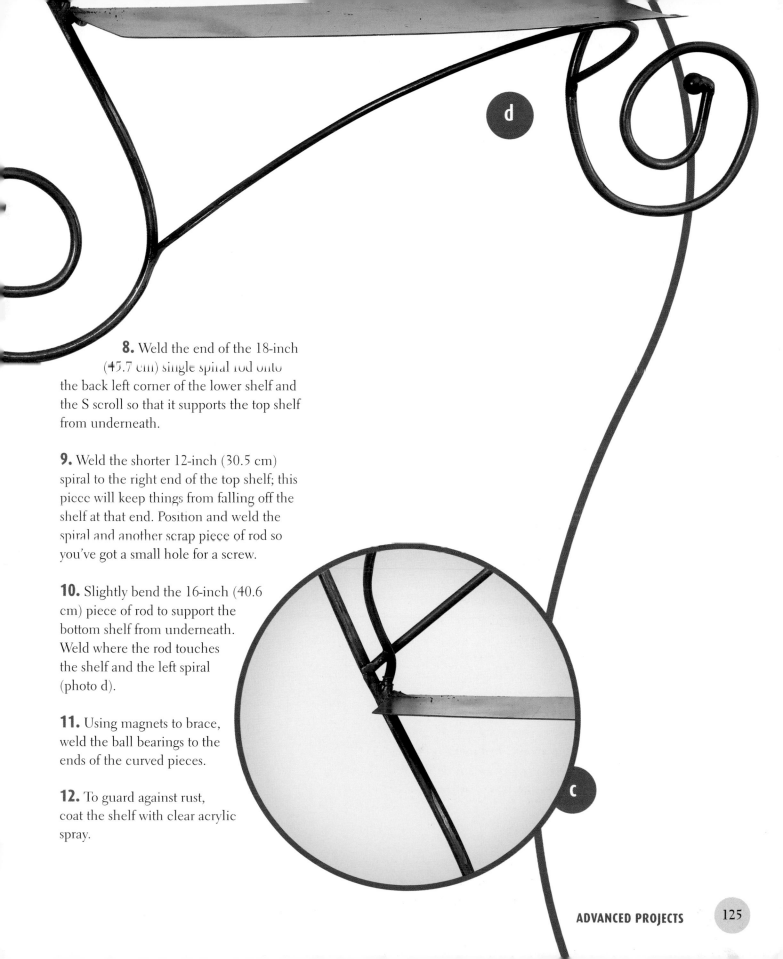

8. Weld the end of the 18-inch (45.7 cm) single spiral rod onto the back left corner of the lower shelf and the S scroll so that it supports the top shelf from underneath.

9. Weld the shorter 12-inch (30.5 cm) spiral to the right end of the top shelf; this piece will keep things from falling off the shelf at that end. Position and weld the spiral and another scrap piece of rod so you've got a small hole for a screw.

10. Slightly bend the 16-inch (40.6 cm) piece of rod to support the bottom shelf from underneath. Weld where the rod touches the shelf and the left spiral (photo d).

11. Using magnets to brace, weld the ball bearings to the ends of the curved pieces.

12. To guard against rust, coat the shelf with clear acrylic spray.

about
the author

Although Kimberli Matin was the "artist in the family," her passion for marketing and product development led her to obtain a communications degree from Portland State University and an early career in marketing and public relations.

It wasn't long, however, before the call to create art became too strong to ignore. After experimenting with various art forms, including a successful painted clothing business, Kimberli began creating with steel in the early 1990s after asking a neighbor to make a mask for her out of an old water heater. Since then she has designed, handcrafted, and sold thousands of pieces to more than eighty galleries throughout the United States, recently inventing a new process for planning and creating pieces by scanning architectural catalog parts and arranging them on the computer.

Kimberli greatly enjoys teaching Welding for Fun workshops at various locations across the country. To view more of her work—and find out about workshops and classes—visit www.weldingforfun.com or go to www.artmetal.com to communicate with Kimberli and other metal artists from around the world. Born and raised in Portland, Oregon, Kimberli currently lives close to nature in the beautiful backwoods of North Carolina.

acknowledgments

I would like to extend a huge thank you to my editor and everyone at Lark Books who nurtured me along during the tough times and provided some laughs along the way. I am extremely grateful for the opportunity to get this new perspective on welding out to the world so beautifully.

Thank you to Vicky Rhine, whose idea it was to contact Lark Books in the first place.

Thank you to Enrique Vega for his constant support and help with technical details as well as the use of his wonderful shop when I needed it.

Thank you to everyone from www.artmetal.com for their helpful suggestions about what should be included.

Thank you to my friends and family who listened to me and supported me throughout this process.

Thank you to Bish Enterprises, Lee Iron & Metal, and D.H. Griffin scrapyards for being kind and supportive to us incessant scrap hounds.

Thank you to Praxair and Airgas National Welders for being such wonderful companies to do business with.

Thank you to Stelphanie Williams and Central Carolina Community College for supporting my first few fledgling classes.

And finally, thank you to the wonderful photographer Steve Mann; to Nicole McConville, who sympathetically helped me through some challenging hurdles; to my art director Kristi Pfeffer; and to Amanda Carestio for answering my many questions and for magically making my instructions understandable.

Index

A Word About Safety

Please exercise caution, common sense, and good judgment when working with the tools and techniques contained in this book. The publisher cannot guarantee the accuracy, adequacy, or the completeness of the information contained in this book and disclaims all warranties, express or implied, regarding the information. The publisher does not assume any responsibility for use of this book, and any use by a reader is at the reader's own risk. The publisher will not be liable for any direct, indirect, consequential, special, exemplary, or other damages arising from use of this book.